Markus Hohl
Andreas Gigon
Philippe Jeanneret

Grasshopper and butterfly diversity in the Swiss Alps

Markus Hohl
Andreas Gigon
Philippe Jeanneret

Grasshopper and butterfly diversity in the Swiss Alps

Spatial and temporal variation of grasshopper and butterfly communities in semi-natural grasslands of the Swiss Alps

Südwestdeutscher Verlag für Hochschulschriften

Impressum/Imprint (nur für Deutschland/only for Germany)
Bibliografische Information der Deutschen Nationalbibliothek: Die Deutsche Nationalbibliothek verzeichnet diese Publikation in der Deutschen Nationalbibliografie; detaillierte bibliografische Daten sind im Internet über http://dnb.d-nb.de abrufbar.
Alle in diesem Buch genannten Marken und Produktnamen unterliegen warenzeichen-, marken- oder patentrechtlichem Schutz bzw. sind Warenzeichen oder eingetragene Warenzeichen der jeweiligen Inhaber. Die Wiedergabe von Marken, Produktnamen, Gebrauchsnamen, Handelsnamen, Warenbezeichnungen u.s.w. in diesem Werk berechtigt auch ohne besondere Kennzeichnung nicht zu der Annahme, dass solche Namen im Sinne der Warenzeichen- und Markenschutzgesetzgebung als frei zu betrachten wären und daher von jedermann benutzt werden dürften.

Coverbild: www.ingimage.com

Verlag: Südwestdeutscher Verlag für Hochschulschriften GmbH & Co. KG
Heinrich-Böcking-Str. 6-8, 66121 Saarbrücken, Deutschland
Telefon +49 681 37 20 271-1, Telefax +49 681 37 20 271-0
Email: info@svh-verlag.de

Approved by: Zurich, Swiss Federal Institute of Technology ETH Zurich, Diss. ETH No.16624, 2006

Herstellung in Deutschland (siehe letzte Seite)
ISBN: 978-3-8381-3345-4

Imprint (only for USA, GB)
Bibliographic information published by the Deutsche Nationalbibliothek: The Deutsche Nationalbibliothek lists this publication in the Deutsche Nationalbibliografie; detailed bibliographic data are available in the Internet at http://dnb.d-nb.de.
Any brand names and product names mentioned in this book are subject to trademark, brand or patent protection and are trademarks or registered trademarks of their respective holders. The use of brand names, product names, common names, trade names, product descriptions etc. even without a particular marking in this works is in no way to be construed to mean that such names may be regarded as unrestricted in respect of trademark and brand protection legislation and could thus be used by anyone.

Cover image: www.ingimage.com

Publisher: Südwestdeutscher Verlag für Hochschulschriften GmbH & Co. KG
Heinrich-Böcking-Str. 6-8, 66121 Saarbrücken, Germany
Phone +49 681 37 20 271-1, Fax +49 681 37 20 271-0
Email: info@svh-verlag.de

Printed in the U.S.A.
Printed in the U.K. by (see last page)
ISBN: 978-3-8381-3345-4

Copyright © 2012 by the author and Südwestdeutscher Verlag für Hochschulschriften GmbH & Co. KG and licensors
All rights reserved. Saarbrücken 2012

Table of contents

Summary		3
Zusammenfassung		5
General Introduction		9
I	Effects of land use and climate change on diurnal Lepidoptera communities in semi-natural grasslands of the central Swiss Alps	15
II	Comparison of grasshopper assemblages recorded in 1981-83 and in 2002-03 as an indication of land use change in Grindelwald, northern Swiss Alps	45
III	Influence of grassland management, altitude and slope on Orthoptera and diurnal Lepidoptera communities in two Swiss Alpine valleys	66
Acknowledgements		97

Recommendation for Citation: Example

M. Hohl, P. Jeanneret, A. Gigon and A. Erhardt (2012) Effects of land use and climate change on diurnal Lepidoptera communities in semi-natural grasslands of the central Swiss Alps. In M. Hohl, A. Gigon and P. Jeanneret (2012) Grasshopper and butterfly diversity in the Swiss Alps, Südwestdeutscher Verlag für Hochschulschriften, Saarbrücken

Summary

In the Swiss Alps, traditionally cultivated grasslands are known to be species rich in Orthoptera and Lepidoptera. Due to the ongoing abandonment or the intensification of management across the Alpine arc these extensively managed grasslands are decreasing and becoming of high conservation interest. Up to now, information has been scarce regarding the consequences of these changes on grasshopper and butterfly diversity in the grasslands of the Alps. Because Orthoptera and Lepidoptera react sensitively to habitat changes, we studied these two taxa in intensively and extensively managed meadows and in lightly grazed pastures in two valleys of the Swiss Alps; Grindelwald (BE) and Tavetsch (Tujetsch, GR).

The study aimed to document changes in grasshopper and butterfly communities over the last three decades using historical data. We wanted to assess potential causes for observed changes and to determine the importance of grassland management for the conservation of grasshoppers and butterflies in the Swiss Alps.

In study I, we investigate changes in diurnal Lepidoptera communities of semi-natural grasslands in the subalpine zone by repeating in 2002-04 surveys made in 1977-79, using the same eleven study sites in the central Swiss Alps (Tavetsch Valley) and the same recording technique.

In 2002-04, 133 species were observed (77 butterflies and 56 diurnal moths), four less than in 1977-79. The average species number per site in the meadows and pastures was stable over the last 25 years, i.e. about 70 species in 1977-79 and in 2002-04. However, the number of butterfly species per site increased, whereas the number of diurnal moth species per site decreased. The species composition of the communities changed significantly between 1977-79 and 2002-04. 31 species showed a statistically significant decrease in abundance, while 15 species showed a significant increase. Lepidoptera restricted to extensively managed grasslands decreased, and species with a wide habitat range increased. This suggests that the grassland management was intensified over the last 25 years. At the same time, subalpine-alpine species experienced losses and lowland species immigrated into the valley, indicating a vertical shift of species into higher elevations, probably due to the effects of global warming.

In study II, we compare grasshopper assemblages (with a total of 21 species) of differently managed grasslands recorded in 1981-83 with records of 2002-03 in the same twenty plots in Grindelwald. The plots were situated between 900 and 2000 m and they were of different steepness.

In 2002-03, all species found in 1981-83 were refound. The difference between the grasshopper assemblages of 1981-83 and 2002-03 were very small, but statistically significant. Some species no longer occurred locally in particular plots in 2002-03, but were found in other plots in 2002-03 when compared to 1981-83. The comparison of the species composition over time indicated no substantial loss in the species assemblages from 1981-83 to 2002-03. This suggests that no general change of land use with detrimental effects on grasshoppers has occurred over the last two decades in the Grindelwald region.

The spatial analysis showed that both altitude and slope had a significant effect on the composition of the species assemblages. The assemblages in the steep plots of the upper montane zone (1100-1500 m) were the most species rich.

In study III, we analyse the influence of the different management types, the altitude and the slope on the present diversity of both taxa in the Grindelwald and Tavetsch valleys.

In total, we observed 28 Orthoptera (without Tetrigidae) and 101 Lepidoptera species of the Rhopalocera, Hesperiidae and Zygaenidae in the two study sites; out of these, 11 Orthoptera and 38 Lepidoptera are currently on the Red List of Northern Switzerland. The average number of Orthoptera species per site did not significantly differ among the management types, while the average number of Lepidoptera species was lowest in the intensively managed meadows.

The high topographic variability and the variety of the management forms in the valleys are a major source of species diversity. The diversity of both insect groups was highest in extensively managed meadows, i.e. 17-20 Orthoptera species and 66-73 Lepidoptera species, and in lightly grazed pastures in which we observed 18-20 Orthoptera species and 63-68 Lepidoptera species.

Implications for Nature Conservation

Our studies confirmed that cultivated grasslands in the Swiss Alps are of high conservation value. Maintaining a variety of extensive grassland management forms is a key for successful conservation of Orthoptera and Lepidoptera in the Alps. Special focus should be given to extensively managed meadows and lightly grazed pastures in the upper montane and subalpine zones. In agri-environmental programmes and conservation plans for grasslands, the habitat requirements of these taxa must be considered in order to prevent further declines, for butterflies in particular. For maintaining the Lepidoptera and Orthoptera diversity a rotational management of cultivated and abandoned grasslands should be taken into account. Large-scale networks of extensively managed grasslands across altitudinal zones are needed to allow species to expand into areas with suitable climatic conditions.

Zusammenfassung

Traditionell bewirtschaftete Wiesen und Weiden in den Schweizer Alpen sind artenreiche Heuschrecken- und Schmetterlingslebensräume. Aufgrund der zunehmenden Verbrachung von Grenzertragsflächen und der Intensivierung der Bewirtschaftung in Gunstlagen, nimmt dieses extensiv genutzte Grasland ab. Deswegen wächst das Interesse des Naturschutzes an solchen Graslandsystemen.

Über die Konsequenzen der Landnutzungsänderungen auf die Heuschrecken- und Schmetterlingsfauna in den Alpen ist bis jetzt nur sehr wenig bekannt. Weil Heuschrecken und Schmetterlinge sensibel auf Veränderungen in ihren Habitaten reagieren, haben wir beide Insektengruppen in intensiv und extensiv bewirtschafteten Wiesen und in extensiv genutzten Weiden untersucht. Ziel der Untersuchung war es, Veränderungen in Heuschrecken- und Schmetterlingsgemeinschaften während der letzten 2-3 Jahrzehnte anhand historischer Daten zu dokumentieren und mögliche Gründe zu ermitteln, die zu den beobachteten Veränderungen geführt haben. Mit den gewonnen Erkenntnissen sollte die Bedeutung von Wiesen und Weiden für den Schmetterlings- und Heuschreckenschutz in den Alpen aufgezeigt werden.

In Untersuchung I haben wir Veränderungen in Gemeinschaften von tagaktiven Lepidopteren im Grasland in der subalpinen Zone untersucht. Dazu haben wir Erhebungen von Andreas Erhardt, die er 1977-79 in Wiesen und Weiden des Tavetsch durchgeführt hatte, 2002-2004 in denselben elf Untersuchungsflächen und mit derselben Methode wiederholt.

2002-04 wurden 133 Arten beobachtet (77 Tagfalter und 56 tagaktive Heterocera) vier Arten weniger als 1977-79. Die durchschnittliche Artenzahl pro Untersuchungsfläche blieb während der letzten 25 Jahre unverändert, d.h. 70 Arten in 1977-79 und 2002-04. Die durchschnittliche Artenzahl von Tagfaltern pro Fläche nahm aber zu, wohingegen die Artenzahl tagaktiver Heterocera abnahm.

Die Zusammensetzung der Lepidopterengemeinschaften hat sich zwischen 1977-79 und 2002-04 signifikant verändert. 31 Arten zeigten eine signifikante Abnahme in der Abundanz, während für 15 Arten eine Zunahme verzeichnet werden konnte. Lepidopteren, die auf extensiv bewirtschaftetes Grasland angewiesen sind, erlitten Verluste, während Arten mit geringen Lebensraumansprüchen zunahmen. Diese Veränderung lässt vermuten, dass die Bewirtschaftung des Graslands in den letzten 25 Jahren intensiviert wurde. Gleichzeitig wurde festgestellt, dass die Abundanz subalpin-alpiner Lepidopteren zurückging und Arten aus dem Tiefland in das Tal eingewandert sind. Die vertikale Verschiebung von Arten in höhere Lagen ist möglicherweise auf die globale Klimaerwärmung zurückzuführen.

In Untersuchung II haben wir Heuschrecken-Artengarnituren von 1981-83 mit solchen von 2002-03 verglichen. Die Aufnahmen wurden in denselben 20 unterschiedlich bewirtschafteten Wiesen und Weiden in Grindelwald wiederholt. Die Flächen lagen zwischen 900 und 2000 m und waren unterschiedlich geneigt.

Alle Arten, die 1981-83 in den Flächen gefunden wurden, konnten 2002-03 noch nachgewiesen werden. Die Unterschiede in den Artengarnituren von 1981-83 und 2002-03 waren klein, aber statistische signifikant. Einige Arten sind aus bestimmten Flächen verschwunden. Diese Arten konnten aber 2002-03 in Flächen nachgewiesen werden, in denen sie 1981-83 nicht beobachtet wurden.

Die geringen Veränderungen in der Zusammensetzung der Artengarnituren zwischen 1981-83 und 2002-03 zeigen keine substantiellen Verluste von Arten an. Dies gibt zur Annahme Anlass, dass sich die Bewirtschaftung des Graslands während der letzten 20 Jahre in Grindelwald kaum zum Nachteil der Heuschreckendiversität verändert hat.

Die räumliche Analyse hat gezeigt, dass sowohl die Höhe, wie auch die Neigung einen signifikanten Einfluss auf die Zusammensetzung der Heuschreckengarnituren hatten. Die steilen Flächen in der obermontanen Stufe (1100-1500) erwiesen sich dabei als die artenreichsten Wiesen und Weiden.

In Untersuchung III haben wir den Einfluss der verschiedenen Nutzungstypen, der Höhe und der Neigung auf die aktuelle Diversität von Heuschrecken und Schmetterlingen in den beiden Tälern Grindelwald und im Tavetsch analysiert.

Im Ganzen fanden wir in den beiden Untersuchungsgebieten 28 Heuschrecken- und 101 Tagfalterarten (Rhopalocera, Hesperiidae, Zygaenidae). Davon sind elf Heuschrecken und 38 Tagfalter in den Roten Listen für die Nordschweiz aufgeführt.

Zwischen den durchschnittlichen Artenzahlen von Heuschrecken in den verschieden Nutzungstypen konnten keine statistisch gesicherten Unterschiede gefunden werden. Bei den Tagfaltern war die durchschnittliche Artenzahl in intensiv genutzten Wiesen aber am tiefsten.

Die topographische Variabilität und die vielfältigen Bewirtschaftungsformen des Graslands tragen wesentlich zur Heuschrecken und Schmetterlingsdiversität in diesen Tälern bei. In extensive genutzten Wiesen und Weiden war die Diversität der beiden Insektengruppen am grössten. In diesen Nutzungstypen wurden jeweils zwischen 17 und 20 Heuschreckenarten bzw. 63-73 Tagfalterarten beobachtet.

Schlussfolgerungen für den Naturschutz

Unsere Untersuchungen haben bestätigt, dass der Naturschutzwert von Wiesen und Weiden in den Alpen sehr gross ist. Die Erhaltung einer extensiven Graslandnutzung spielt beim Schutz von Heuschrecken und Schmetterlingen im Berggebiet eine zentrale Rolle. Von besonderer Bedeutung sind die extensiv genutzten Wiesen und Weiden in der montanen und der subalpinen Stufe. Um weitere Verluste dieser Insekten, insbesondere von Tagfaltern, zu vermeiden, müssen ihre Lebensraumansprüche in den Bewirtschaftungsauflagen des Agrarökologischen Programms und in Naturschutzmassnahmen berücksichtigt werden. Um die Schmetterlings- und Heuschreckendiversität in den Tälern zu erhalten, sollte ein Rotationsmanagement von bewirtschaftetem Grasland und kurzzeitigen Brachen in Betracht gezogen werden. Zusätzlich sind grossräumige Netzwerke von extensiv genutztem Grasland zwischen den verschiedenen Höhenstufen nötig, damit sich die Arten in klimatisch günstige Lagen ausbreiten können.

General Introduction

Over the last century, profound changes in land use have been observed in the Alps due to changes in socio-economical conditions. Mainly in the large valleys and in tourist centres land use has been increasingly intensified (Internationale Alpenschutzkomission CIPRA, 2001). At the same time, the depopulation of villages in far reaches has further progressed and side valleys have been abandoned. These changes have been sped up by globalization, European integration and the liberalization of markets (Internationale Alpenschutzkomission CIPRA, 2001).

Beside land use changes, also changes in the climatic conditions have started to attract increased attention in the Alps (e.g. Internationale Alpenschutzkomission CIPRA, 2001; Körner & Spehn 2002; Bätzing 2003). The massive retreat of the Alpine glaciers and the defrosting of permafrost soils are worrying examples of the ongoing effects of global warming occurring in this ecosystem.

About 15% of the whole Alpine Arc is located in Switzerland and covers about 65% of the total surface of the country (Bätzing 1997). Thus the Alps represent an important landscape and ecosystem for Switzerland. In regard to the ongoing changes, the question arises: what is a sustainable development for the Alpine region of Switzerland, and what is needed to set it in motion (Schweizerischer Nationalfonds 2003).

Following these considerations, the National Research Programme NRP48 called *'Landscapes and habitats of the Alps'* was initialised. This programme aims to determine in an interdisciplinary approach which developments in the Alpine area are discernible, socially desirable, ecologically sound, and economically sustainable (Schweizerischer Nationalfonds 2003).

The present study was carried out in the frame of the NRP48 and was aimed to document the present biodiversity of cultivated grasslands in the Swiss Alps (Lüscher *et al.* 2003).

Cultivated grasslands are characteristic landscape elements and represent the major form of agricultural land use in the Swiss Alps. At present, about 400'000 ha are cultivated (Alpine pastures not included!): 88% are typical grassland systems, i.e. meadows and pastures (Internationale Alpenschutzkomission CIPRA 2001). These grasslands were all created by human activity over the last millennia (Ellenberg 1996). Until recently, they were extensively managed due to the difficult land use conditions in the Alps (Bätzing 2003). The natural dynamic, the rough climatic conditions and the short vegetation periods limit the management period and its intensity, and hence the yields. Over the last centuries, this extensive management created semi-natural grasslands of high species diversity and unique plant species composition (e.g. Bischof 1984; Pfister 1984; Ellenberg 1996).

However, traditionally cultivated grasslands are experiencing similar losses in the Alps (Internationale Alpenschutzkomission CIPRA 2001) as in other landscapes of Europe (e.g. Poschlod & WallisDeVries 2002; Strijker 2005). Due to the growing economical pressure on agriculture the farmers are forced to rationalize their traditional management forms. In many places, meadows and pastures easy to cultivate are intensified, i.e. more fertiliser is applied, more frequent mowing occurs and livestock pressure is larger than in the past. At the same time, grasslands difficult to access and to manage become abandoned and turn into forests in the long term (Bischof 1984). Both developments cause a loss of grassland diversity (Bischof 1984; Erhardt 1985; Dietl 1996; Balmer & Erhardt 2000).

Our study focused on grasshoppers and butterflies because meadows and pastures are important habitats for both insect groups (e.g. Kiser 1981; Erhardt 1985, b; Schiess 1988; Wettstein & Schmid 1999; Oertli et al. 2005). In the meadows and pastures, grasshoppers and butterflies find food, host plants, mating places and the required opportunities for reproduction and hibernation (Detzel 1998; Ingrisch & Köhler 1998; Lepidopterologen-Arbeitsgruppe 1991). Compared to the Swiss lowlands where many grasshopper and butterfly populations have been lost or shrank to small, isolated and thus vulnerable rest populations (Gonseth 1987; Nadig & Thorens 1997), the diversity of the two taxa is still high in the Swiss Alps (e.g. Erhardt 1985, Schiess 1988; Erhardt 1995; Lepidopterologen-Arbeitsgruppe 1991).

However, due to the observed land use and climate changes, impacts on the diversity of these insects must be expected because grasshoppers and butterflies are known to react sensitively to habitat changes due to their specific ecological requirements (Erhardt 1985; Detzel 1998; Ingrisch & Köhler 1998).

Up to now, only little is known about changes to grasshopper and butterfly communities in the Swiss Alps over the last decades because monitoring programmes on insects started only recently (see Hintermann et al. 2002). Therefore, new collected data have to be compared to historical data to investigate the changes in biological communities over more than a few years and for establishing required monitoring programmes (see Schiess & Schiess-Bühler 1997: Shaffer et al. 1998).

In our study, we aimed to document the changes in the species composition of Orthoptera and Lepidoptera communities in semi-natural grassland of two valleys in the Swiss Alps over the last 20-30 years using historical records. Furthermore, we analysed and discussed the importance of the grassland management for grasshopper and butterfly conservation in the Alps.

Butterfly and grasshopper communities were investigated in meadows and pastures of the Grindelwald region (BE) and the Tavetsch valley (Tujetsch, GR) and were compared with historical records collected more than 20 years ago.

In chapter I, butterfly surveys in subalpine meadows and pastures, carried out in 1977-79 by A. Erhardt (1985) in the Tavetsch valley, are compared to surveys made in 2002-04. Changes in species numbers, the increase and decrease of individual species as well as changes in species compositions are addressed. The changes are discussed in the context of the observed land use changes in the valley and probable effects of global warming.

In chapter II, we compare the grasshopper assemblages recorded in 2002-03 with those observed by H. Schiess (1988) in 1981-83 in 20 meadows and pastures of Grindelwald. Both, the loss and the gain of species in the assemblages over the last 20 years are assessed and are analysed with regard to different altitudinal zones and the slope.

In chapter III, we investigate the influence of different management types, the altitude and the slope on the present Orthoptera and Lepidoptera diversity on the south slopes of Grindelwald and Tavetsch. We analyse differences among the species communities of intensively and extensively mown meadows and of pastures in a heterogeneous, mosaic-like landscape and discuss the current conservation value of grassland habitats in these areas.

The study was closely connected to theses of M. Peter (2006) and D. Kampmann (in prep.), which focused their investigations on the vegetation and Orthoptera diversity in cultivated grassland of several regions in the Alps.

References

Balmer, O. and Erhardt, A. 2000. Consequences of succession on extensively grazed grasslands for central European butterfly communities: Rethinking conservation practices. Conservation Biology 14: 746-757.

Bätzing, W. 1997. Kleines Alpenlexikon: Umwelt, Wirtschaft, Kultur. Beck'sche Reihe 1205, München.

Bätzing, W. 2003. Die Alpen: Geschichte und Zukunft einer europäischen Kulturlandschaft. C.H. Beck, München.

Bischof, N. 1984. Pflanzensoziologische Untersuchungen von Sukzessionen aus gemähten Magerrasen in der subalpinen Stufe der Zentralalpen. Beiträge zur geobotanischen Landesaufnahme der Schweiz 60: 1-128.

Detzel, P. 1998. Die Heuschrecken Baden-Württembergs. Ulmer, Stuttgart.

Dietl, W. 1996. Zur pfleglichen Nutzung der Wiesen und Weiden im Berggebiet. Montagna 6: 17-18.

Ellenberg, H. 1996. Vegetation Mitteleuropas mit den Alpen in ökologischer, dynamischer und historischer Sicht, 5th ed. Ulmer, Stuttgart.

Erhardt, A. 1985. Wiesen und Brachland als Lebensraum für Schmetterlinge. Birkhäuser, Basel.

Erhardt, A. 1995. Ecology and conservation of alpine Lepidoptera. Pages 258-276 in Pullin A. S., editor. Ecology and conservation of butterflies. Chapman & Hall, London.

Hintermann, U., Weber, D., Zangger, A., Schmill, J. (2002). Biodiversity Monitoring in Switzerland, BDM-Interim Report, Swiss Agency for Environment, Forests and Landscapes SAEFL, 342: 1- 89.

Ingrisch, S. and Köhler, G. 1998. Die Heuschrecken Mitteleuropas, Westarp Wissenschaften, Magdeburg.

Internationale Alpenschutzkomission CIPRA, 2001. 2 Alpenreport: Daten. Fakten, Probleme, Lösungsansätze. Paul Haupt, Bern, Stuttgart, Wien.

Kampmann, D. In prep. Dissertation an der Universität Freiburg, Freiburg.

Kiser, K. 1987. Tagaktive Grossschmetterlinge als Bioindikatoren für landwirtschaftliche Nutzflächen der Zentralschweizer Voralpen. Dissertation, Universität Freiburg Schweiz, Freiburg.

Körner, C. and Spehn, E. 2002. Mountain biodiversity: a global assessment. Parthenon, New York, London.

Lepidopterologen-Arbeitsgruppe 1991. Tagfalter und ihre Lebensräume. Vol. 1, Schweizerischer Bund für Naturschutz, Basel.

Lüscher, A., Hohl, M., Kampmann, D., Peter, M., Herzog, F. and Jeanneret, P. 2003. Driving forces for changes in management and biodiversity of Alpine grasslands – basis for planning future development. Progress Meeting des NFP48, Progress Report, Bern.

Oertli, S., Müller, A., Steiner, D., Breitenstein, A. and Dorn, S. 2005. Cross-taxon congruence of species diversity and community similarity among three taxa in a mosaic landscape. Biological Conservation 126: 195-205.

Peter, M. 2006. Long term floristic changes of permanent grasslands of the Swiss Alps. Dissertation ETH Zürich, Zürich.

Pfister, H. 1984. Grünlandgesellschaften, Pflanzenstandort und futterbauliche Nutzungsvarianten im montan-subalpinen Bereich. Schlussbericht des Schweizerischen MAB-Programm Nr. 7, Bundesamt für Umweltschutz, Bern.

Poschlod, P. and WallisDeVries, M. F. 2002. The historical and socioeconomical perspectives of calcareous grasslands – lessons from the distant and recent past. Biological Conservation 104: 361-376.

Schiess, H. 1988. Wildtiere in der Kulturlandschaft. Schlussbericht des Schweizerischen MAB-Programm Nr. 35, Bundesamt für Umweltschutz, Bern.

Schiess, H. and Schiess-Bühler, C. 1997. Dominanzminderung als ökologisches Prinzip: Eine Neubewertung der ursprünglichen Waldnutzungen für den Arten- und Biotopschutz am

Beispiel der Tagfalterfauna eines Auenwaldes in der Nordschweiz. Mitteilungen der Eidgenössischen Forschungsanstalt für Wald, Schnee und Landschaft, Band 72, Birmensdorf.

Schweizerischer Nationalfonds 2003. Landscape and habitats of the Alps: Portrait of the National Research Programme NRP 48, Bern.

Shaffer, H. B., Fisher, R. N. and Davidson, C. 1998. The role of natural history collections in documenting species declines. Trends in Ecology and Evolution 13: 27-30.

Strijker, D. 2005. Marginal lands in Europe – causes of decline. Basic and Applied Ecology 6: 99-106.

Wettstein, W. and Schmid, B. 1999. Conservation of arthropod diversity in montane wetlands: effect of altitude, habitat quality and habitat fragmentation on butterflies and grasshoppers. Journal of Applied Ecology 36: 363-373.

I. Effects of land use and climate change on diurnal Lepidoptera communities in semi-natural grasslands of the central Swiss Alps

M. Hohl[1], P. Jeanneret[1], A. Gigon[2] and A. Erhardt[3]

[1]Agroscope FAL Reckenholz, Swiss Federal Research Station for Agroecology and Agriculture, Reckenholzstr. 191, CH-8046 Zurich, Switzerland

[2]Institute of Integrative Biology, Swiss Federal Institute of Technology, ETH Zurich, CH-8092 Zurich, Switzerland

[3]Department of Integrative Biology, University of Basel, St. Johanns-Vorstadt 10, CH-4056 Basel, Switzerland

Abstract

In the subalpine zone of the Swiss Alps, traditionally cultivated, semi-natural grasslands are habitats rich in diurnal Lepidoptera and of high conservation value. However, over the last decades, the environmental and socio-economic conditions have changed profoundly. Up to now, information has been scarce regarding the consequences of these changes on the Lepidoptera in these habitats. Lepidoptera react sensitively to habitat changes. They are recognized as bioindicators and are well known as representatives for other insects. In order to document potential changes in Lepidoptera communities on semi-natural grasslands of the subalpine zone, we repeated Lepidoptera surveys of 1977-1979 in 2002-2004, using the same eleven study sites in the central Swiss Alps (Tavetsch Valley) and the same recording technique.

In 2002-04, 133 species (77 butterflies and 56 diurnal moths) were recorded, four less than in 1977-79. The average species number per site in the former meadows and pastures was stable over the last 25 years, i.e. about 70 species in 1977-79 and in 2002-04. However, the number of butterfly species per site increased, whereas the number of diurnal moth species per site decreased.

The species composition of the communities changed significantly between 1977-79 and 2002-04. 31 species showed a statistically significant decrease while 15 species showed a significant increase in abundance. Species restricted to extensively managed grasslands decreased, and species with a wide habitat range increased. This suggests that the grassland management was intensified over the last 25 years. At the same time, subalpine-alpine species experienced losses and lowland species immigrated into the valley, indicating a vertical shift of species into higher elevations, probably due to the effects of global warming.

These major changes in the butterfly communities of the Tavetsch Valley are a cause for concern as they show a trivialisation in Lepidoptera communities caused by habitat degradation and global warming.

In agri-environmental programmes and conservation plans for grasslands, the habitat requirements of butterflies and other insects must be considered in order to prevent further decline. Large-scale networks of extensively managed grasslands across altitudinal zones are required to allow species to expand into areas with suitable climatic conditions.

Keywords: Lepidoptera, biodiversity, grassland management, abandonment, global warming, subalpine zone, Alps

South slope of the Tavetsch Vally (Grisons, Switzerland)

Introduction

Traditionally cultivated grasslands belong to the most species rich habitats in the subalpine zone of the Swiss Alps. In addition to their high floristic diversity (Bischof 1981), Erhardt (1985a) showed 25 years ago that these lightly cultivated meadows and pastures are very species rich Lepidoptera habitats in the Alps. Because a variety of rare and typical mountain Lepidoptera occurs in these grasslands, they are of high conservation value (Erhardt, 1985a, b). However, we have no information how the Lepidoptera communities of these grasslands have changed over the last decades.

During the first decades after the Second World War, changes of the traditional grassland management were observed due to changes in the socio-economic and political conditions of mountain agriculture in Switzerland (Erhardt 1985a, b). On the one hand, large parts of grasslands in the Swiss Alps, in particular lightly mown meadows or lightly grazed grasslands on steep slopes, which are difficult to cultivate, were abandoned. On the other hand, the agricultural management on areas easier to manage was intensified. It must be expected that these changes have continued over the last two decades because the demand for higher yields and the price of agricultural labour have further increased. Furthermore, shrub encroachment and natural reafforestation probably continued in the grasslands already abandoned in the seventies.

For maintaining semi-natural grasslands, an extensive management, i.e. mowing and/or grazing is necessary. If these grasslands become abandoned, they turn into climax vegetation, i.e. forests in the long-term (Bischof 1984). Consequently, the species-rich Lepidoptera communities of these grasslands would be replaced by characteristic, but species-poor Lepidoptera communities of forests (Erhardt 1985b). If, on the other hand, the management of these habitats is intensified, their Lepidoptera communities become trivialized because more frequent mowing, grazing, and fertiliser application harm the larval stages and change the plant species composition of host plants and nectar resources (Erhardt 1985b).

Over the last decade, several authors have reported that butterflies and other taxa like birds are responding to climate warming and are shifting their distribution to track the current temperature increase (Parmesan *et al.* 1999; Thomas & Lennon 1999; Warren *et al.* 2001; Hill *et al.* 2002; Konvicka *et al.* 2003; Parmesan & Yohe 2003). High elevations in the Alps are expected to experience above average warming with continued global warming (Guisan *et al.* 1995; Theurillat & Guisan 2001; Beniston 2003; Burga *et al.* 2003). There is already evidence that changes in high mountain summit flora in the Alps reflect effects of climate change (Grabherr *et al.* 1994; Walther *et al.* 2005). Thus, beside the effects of a changing grassland management, that climate warming might cause also additional changes in the composition of Lepidoptera communities in the Alps.

Lepidoptera react sensitively on changes to their environment due to their specific habitat requirements (Erhardt 1985a; Erhardt & Thomas 1991). They are well known bioindicators and are representative for other insects (Ehrlich 1994; Master *et al.* 2000). Monitoring studies in Britain have shown in recent decades that butterflies have experienced greater extinction rates than birds or plants (Thomas *et al.* 2004). This stresses the importance of monitoring butterflies in these semi-natural habitats.

For studying and assessing the development of Lepidoptera communities in the subalpine grasslands of the Swiss Alps under the premise of a changing agricultural land use and climate warming, we repeated the Lepidoptera surveys of 1977-79 carried out by Erhardt (1985a, b) in 2002-04 on the same sites in the subalpine zone of the Swiss Alps.

The aim of the study was to document changes in subalpine Lepidoptera communities between 1977-79 and 2002-04 and to deduce conservation recommendations for maintaining these insects and grassland diversity in general. We asked (1) how the Lepidoptera diversity of subalpine grassland has developed over the last 25 years, (2) which species are showing increases and which species experienced losses and (3) whether the observed changes in species composition indicate habitat degradation and effects of global warming.

Study area and methods

Investigation area and study sites

The study was carried out in the Tavetsch (Grisons), a subalpine valley in the central Swiss Alps. Butterfly surveys on eleven agriculturally managed sites surveyed in 1977-79, were repeated in 2002-04 (Table 1). The sites were located at elevations between 1340 m and 1760 m a.s.l., eight south-facing and three north-facing, and all had similar slope. The size of the investigated sites was equal for 1977-79 and 2002-04 and was 2500 m^2 on average (Erhardt 1985a).

According to Erhardt (1985b), two of the eleven grasslands situated on the south facing slope of the valley, were fertilised and mown once to twice a year; they belonged to the plant association of the Trisetetum flavescentis (Hartmann 1976). Five grasslands were unfertilised hay meadows, which were mown every second year. Among them, four meadows, situated on the south-facing slope belonged to the Polygalo-Poetum violaceae plant association. One meadow situated on the north-facing slope was characterised as a Geo montani-Nardetum maianthemetosum bifoliae (Bischof 1981). Four grasslands were extensively managed pastures, i.e. they were grazed for a short time once in spring and a second time in autumn. Two of them were situated on the north-facing slope and two on the south-facing slope. The plant associations of these pastures were not determined in 1977-79.

Table 1. List of monitored sites, their Swiss grid position, altitude, exposure and the management in 1977-79 and in 2002-04.

Study site	Swiss grid		Altitude	Exposure	Management	
	x	y	m a.s.l		1977-79	2002-04
F1	702716	171071	1460	south	1 x mown per season, fertilised	grazed
F2	698235	168605	1560	south	2 x mown per season, fertilised	grazed
M1	697926	168398	1560	south	1 x mown every second season, unfertilised	grazed
M2	702369	169821	1570	north	1 x mown every second season, unfertilised	1 x mown per season, unfertilised
M3	700223	170662	1570	south	1 x mown every second season, unfertilised	1 x mown per season, unfertilised
M4	698346	168923	1725	south	1 x mown every second season, unfertilised	1 x mown per season, unfertilised
M5	696446	167820	1760	south	1 x mown every second season, unfertilised	lightly grazed
P1	704656	170313	1340	north	lightly grazed	abandoned since 3-5 yr
P2	702284	171123	1480	south	lightly grazed	unchanged
P3	702386	171132	1490	south	lightly grazed	unchanged
P4	697356	167389	1590	north	lightly grazed	abandoned since 3-5 yr

In the last decade, the management practice of nine sites has changed (Table 1). The two fertilised and two unfertilised hay meadows were grazed for about ten years by sheep. The biennial mowing regime of the three other unfertilised hay meadows changed into an annual one since about 1992. The two pastures on the north facing slope were abandoned between 1990 and 2001.

In 2002-04, the vegetation of all the sites was still dominated by grasses and small herbs. In the still managed sites, a few woody plants typically occurring in fallow grasslands are present. Individuals of *Picea abies* (up to 3 m high) occured in most of the grasslands, together with *Betula pendula* and *Rosa canina* at lower altitudes resp. *Alnus viridis* at higher elevations (plant names are according to Lauber & Wagner 2003).

Survey method

The butterfly surveys carried out in the first study period in 1977-79 between June and the end of September were repeated in 2002-04 during the same months. This period is optimal for butterfly recordings at these elevations (Erhardt 1985a). In the extraordinarily hot summer of 2003 the surveys were started earlier, i.e. at the end of May, because activity of spring species was observed earlier that year.

On all sites, the surveys were carried out every 2-3 weeks. Details of the method used to record butterfly data are given in Erhardt (1985a). In summary, an area transect was conducted, i.e. the sites were patrolled in a serpentine pattern and butterflies were identified and counted within a corridor of 5 m. Multiple counting of individuals of fast flying species cannot be completely avoided using this method. However, the resulting error is assumed to be the same in both study periods. Transects were carried out between 10.00 am and 5.00 pm under weather conditions defined by Erhardt (1985a). A single transect took about 20 minutes on average.

In order to repeat the transects of 1977-79 as similarly as possible in 2002-04, some transects were carried out with Andreas Erhardt at the beginning of the data collection in 2002-04.

All Rhopalocera, Hesperiidae and Zygaenidae, as well as diurnal Geometridae, Noctuidae, Arctiidae and Lasiocampidae were recorded. Due to difficulties in distinguishing very similar butterfly species, some species had to be pooled and were considered as one species in statistical analyses (*Boloria napaea/pales, Colias alfacariensis/hyale, Erebia meolans/oeme, Speyeria aglaia/Fabriciana niobe, Distroma trunctata/citrata, Adscita alpina/geryon*).

Nomenclature fallows Leraut (1997).

Classification of the butterfly abundances

In 1977-79, the numbers of counted individuals per species in each surveyed year were added up and transformed into a semi-quantitative scale of 7 classes (Table 2). This transformation already applied by Erhardt (1985a) eliminates small, ecologically less meaningful differences among the species abundance in the different sites (Erhardt, 1985a). In 2002-04, the same procedure was applied.

For the statistical analyses, the average class of abundance per species of the three investigated years in 1977-79 and 2002-04 was calculated and compared on a plot-by-plot basis.

Species observed only with a single individual either in 1977-79 or in 2002-04 were not considered in the statistical analyses.

Table 2. Classes of abundance according to Erhardt (1985a). Flight period = Recording time of one year between Mai and September; Maximum appearance = number of individuals counted in a single transect at one date.

Abundance classes		Number of individuals per study site
1	very rare	1 individual per flight period
2	rare	2-4 individuals per flight period
3	dispersed	5-10 individuals per flight period
4	quite abundant	> 10 individuals per flight period, max. appearance <10 individuals
5	abundant	max. appearance 10-40 individuals
6	very abundant	max. appearance 41-100 individuals
7	extremly common	max. appearance > 100 individuals

Grouping of the species according to their habitats and altitudinal preference

The butterfly species were grouped according to their grassland specialisation and altitudinal preference based on the studies by Erhardt (1985b), Lepidopterologen-Arbeitsgruppe (1991-2000), Forster & Wohlfahrt (1954-1981), Gonseth (1987), Ebert (1991-2003) and Tolman & Lewington (1997).

Grassland specialisation

Stenotopic grassland species (S): These species occur exclusively in extensively managed grasslands such as unfertilised hay meadows and pastures and in the very early stages of abandonment of these habitats. Their larval host plants typically occur in these habitats and the larval stages of these butterflies benefit from the low management pressure.

Eurytopic species (E): These species occur in a wide range of habitats, but they are not dependent on extensively managed grasslands, i.e. they occur in more intensively managed grasslands with trivialised vegetation or prefer later succession stages and woodlands. Their host plants are ubiquitous or do not occur in grasslands.

Altitudinal preference

Colline-montane species (CM): The range of these species is mainly the colline-montane zone. The upper limit of their occurrence is the subalpine zone.

Subalpine-alpine species (SA): These species occur mainly in the subalpine-alpine zone. They do not occur below the border of the upper montane zone. *Palaeochrysophanus hippothoe* was classified as a subalpine-alpine species because in the investigation area, only the subspecies of higher elevations occurs, i.e. *P. hippothoe eurydame.*

Only those species for which we could make a clear classification according to their altitudinal preference were classified in colline-montane or subalpine-alpine species.

Statistical analyses

The change in butterfly communities was analysed by both univariate and multivariate methods.

The multivariate approach is attractive in that it provides an overall significance test of community change. Thus, it avoids the problem of multiple testing in univariate statistics (Ter Braak & Wiertz 1994). Redundancy Analysis (RDA) and partial RDA were used for the analyses of spatial explanatory variables (plot, altitude and exposure) and time (1977-79 and 2002-04) (Ter Braak & Prentice 1988). Monte Carlo permutations were used to test for significance. For each test, 499 permutations were performed. Principal Component Analysis (PCA) was used to visualise the change in the butterfly communities. The multivariate analyses were carried out using the program CANOCO 4.5 (Ter Braak & Šmilauer 2002).

The increase and decrease of species numbers and the species abundance was tested with Wilcoxon signed rank tests (Sokal & Rohlf 1981) using the program S-Plus 6.1®.

Once the overall change was analysed with RDA and partial RDA, the Wilcoxon test helped to single out which species had significantly changed in abundance between 1977-79 and 2002-04. The species' significance was used as a criterion for its display in the ordination diagrams (Ter Braak & Wiertz, 1994).

The change in the number of occupied sites and the average class of abundance between 1977-79 and 2002-04 in the grassland specialisation and the altitudinal preference groups was analysed again using Wilcoxon tests.

Proportions between categorised species were tested using Fishers exact test (Sokal & Rohlf 1981).

Results

Species list of Lepidoptera in 1977-79 and in 2002-04

In total, 148 diurnal Lepidoptera species were found in the 11 study sites in 1977-79 and 2002-04. In 2002-04, 133 species were recorded in the grassland of the Tavetsch, four less than in 1977-79 (Table 3). 122 species of 1977-79 could still be detected in 2002-04.

A simple comparison between the species list of 1977-79 and 2002-04 indicated that changes in the Lepidoptera assemblage of the Tavetsch valley have occurred. After 25 years, 15 species were no longer found, but 11 species were observed only in 2002-04.

In the group of Rhopalocera & Hesperiidae, to which more than half of the species belong, five more butterflies species were found in 2002-04. On the other hand, in the group of diurnal moths, nine species less were found in 2002-04 than in 1977-79.

Table 3. Total number of diurnal Lepidoptera and number of species of different Lepidoptera families in cultivated grassland of the Tavetsch valley and the mean number of species per site in 1977-79 and in 2002-04. The changes in the number of species per site from 1977-79 to 2002-04 was analysed using Wilcoxon tests.

	Total number of species				Mean number of species per sites		
	1977-79	2002-04	Not found in 2002-04	Newly found 2002-04	1977-79	2002-04	Change over time
Rhopalocera & Hesperiidae	72	77	4	9	37.4	47.5	+10.1 **
Papillionidae	2	2	0	0	1.2	1.5	+0.4 n.s
Pieridae	12	12	0	0	6.5	8.7	+2.3 *
Nymphalidae	16	19	2	5	7.4	10.8	+3.5 **
Satyridae	17	18	1	2	7.4	9.2	+1.8 *
Lycaenidae	18	18	1	1	10.9	11.4	+0.5 n.s
Hesperiidae	7	8	0	1	4.1	5.8	+1.7 **
Diurnal Heterocera	65	56	11	2	32.7	23.2	-9.5 **
Geometridae	36	30	6	0	18.5	13.4	-5.10 *
Arctiidae	5	3	2	0	2.1	1	-1.10 n.s
Lasiocampidae	2	2	0	0	1.4	1	-0.40 n.s
Noctuidae	14	13	3	2	5.6	4.3	-1.30 n.s
Zygaenidae	8	8	0	0	5.1	3.5	-1.60 **
All species	137	133	15	11	70.1	70.8	+0.7 n.s

**: $p < 0.01$; *: $p < 0.05$; n.s: not significant

Species number per site in 1977-79 and in 2002-04

Although the total number of species per site has not changed between 1977-79 and 2002-04, i.e. 70.1 in 1977-79 and 70.8 in 2002-04 on average, important differences occured within some butterfly and diurnal moth families (Table 3).

In the Rhopalocera & Hesperiidae, on average, ten more species per site were observed in 2002-04 than in 1977-79, mainly in the families of Pieridae, Nymphalidae, Satyridae and Hesperiidae.

On the other hand, in the diurnal moths between nine and ten species less per site were found in 2002-04 compared to 1977-79 on average, mainly in the families of Geometridae and Zygaenidae.

Increasing and decreasing species abundance

Out of 148 species, twelve species had pair-wise to be pooled and were considered as six species (see methods chapter). Each pair was considered as one species in statistical tests. Out of these 142 species, 31 species decreased and 15 species increased significantly in abundance between 1977-79 and 2002-04 according to the Wilcoxon test (Table 4).

Increases were exclusively observed in the Rhopalocera & Hesperiidae. In this group, 15 species increased and 13 species decreased. In the other investigated families only declines were observed. Nine of 36 Geometridae, one of 2 Lasiocampidae, four of 14 Noctuidae and five of eight Zygaenidae experienced losses.

Out of the 15 species that were no longer found in 2002-04, a statistically confirmed loss was found only for *Boloria napaea/pales* and the moth *Lygris populata*. The other 13 species occurred only locally in 1977-79 and thus no meaningful statistics could be applied. For the same reason, out of the newly occurring nine species, only those of *Inachis io* and *Ochlodes venatus* could be confirmed with statistical significance.

Spatial variation and temporal changes of the butterfly communities

The spatial variation and temporal changes of the 22 butterfly communities (11 sites in 1977-79 + 11 sites in 2002-04) are shown in Figure 1. The first axis represents mainly exposure i.e. the differences between the communities on the south and north-facing slope. The second axis displays mainly the variation of species composition along the altitudinal gradient.

As indicated by arrows in Figure 1, in all study sites changes occurred from 1977-79 to 2002-04. The amount of the temporal change is expressed by the length of these arrows.

In the sites on the north-facing slope (P1, P4, M2) changes were smaller than on the south facing slope, except for the two formerly fertilised meadows (F1, F2). In these two south slope sites the changes were smallest compared to those observed in the other south slopes. The observed changes over time are clearly related to altitude. The arrows, which are connecting corresponding communities of 1977-79 and 2002-04, point in the opposite direction of the altitudinal gradient. Thus the composition of each community of 2002-04 bears a higher resemblance to former communities at lower elevations than to its corresponding community of 1977-79.

Table 4. List of species significantly increasing ($^+$ = p < 0.05; $^{++}$ = p < 0.01) and decreasing ($^-$ = p < 0.05; $^{--}$ = p< 0.01) according to the Wilcoxon test from 1977-79 to 2002-04. Shown are the number of occupied sites and the mean class of abundance in 1977-79 and in 2002-04. E = eurytopic species; S = stenotopic species; CM = colline-montane preference; SA = subalpine-alpine preference.

	Groups		Number of occupied sites			Mean class of abundance		
	Grassland specialisation	Altitudinal preference	1977-79	2002-04	Change	1977-79	2002-04	Change
Increasing species								
Rhopalocera & Hesperiidae								
$^{++}$ *Aporia crataegi*	E		3	11	+8	0.3	1.5	+1.2
$^{++}$ *Pieris brassicae*	E	CM	6	10	+4	0.5	1.2	+0.6
$^{+}$ *Pieris napi*	E	CM	1	6	+5	0.1	0.5	+0.5
$^{++}$ *Leptidea sinapis*	E	CM	1	11	+10	0.1	1.2	+1.1
$^{+}$ *Aglais urticae*	E		10	11	+1	1.7	2.5	+0.7
$^{++}$ *Cynthia cardui*	E		5	11	+6	0.6	1.8	+1.2
$^{++}$ *Inachis io*	E	CM	0	10	+10	0.0	0.9	+0.9
$^{++}$ *Clossiana dia*	S	CM	2	9	+7	0.2	1.8	+1.6
$^{+}$ *Erebia ligea*	E		1	6	+5	0.3	0.9	+0.6
$^{++}$ *Erebia aethiops*	E		2	10	+8	0.4	1.5	+1.2
$^{+}$ *Aphantopus hyperantus*	E	CM	2	6	+4	0.2	1.0	+0.8
$^{++}$ *Polyommatus bellargus*	S		7	10	+3	0.9	1.7	+0.8
$^{+}$ *Lycaena phlaeas*	E	CM	5	10	+5	1.1	2.0	+0.9
$^{++}$ *Thymelicus sylvestris*	E		3	11	+8	0.5	1.7	+1.3
$^{+}$ *Ochlodes venatus*	E	CM	0	5	+5	0.0	0.6	+0.6
Decreasing species								
Rhopalocera & Hesperiidae								
$^{--}$ *Colias phicomone*	S	SA	10	6	-4	2.6	0.6	-2.0
$^{-}$ *Boloria napaea/pales*	S	SA	5	0	-5	0.6	0.0	-0.6
$^{--}$ *Coenonympha gardetta*	S	SA	11	11	0	3.8	2.6	-1.2
$^{--}$ *Erebia melampus*	E	SA	11	11	0	3.8	2.5	-1.3
$^{--}$ *Erebia euryale adyte*	E	SA	9	9	0	3.0	1.5	-1.5
$^{-}$ *Erebia tyndarus*	E	SA	6	4	-2	1.6	0.6	-1.0
$^{-}$ *Lysandra coridon*	S		11	10	-1	4.2	3.3	-0.9
$^{--}$ *Cyaniris semiargus*	S		11	11	0	2.5	1.2	-1.3
$^{--}$ *Palaeochrysophanus hippothoe eurydame*	E	SA	10	10	0	2.5	1.2	-1.3
$^{-}$ *Aricia artaxerxes*	S	SA	10	7	-3	2.1	1.3	-0.8
$^{-}$ *Hesperia comma*	S		10	11	1	2.5	1.5	-1.0
$^{--}$ *Pyrgus alveus*	S		10	6	-4	2.2	0.5	-1.6
Geometridae								
$^{--}$ *Itame flaveolaria*	S	SA	9	8	-1	2.8	1.5	-1.4
$^{-}$ *Crocota pseudotinctaria*	S	SA	9	8	-1	2.5	1.3	-1.3
$^{--}$ *Epirrhoe tristata*	S		10	5	-5	2.5	0.5	-2.0
$^{--}$ *Perizoma albulata*	E		11	7	-4	2.3	0.8	-1.5
$^{--}$ *Xanthoroe montanata*	E		10	4	-6	1.6	0.7	-0.9
$^{--}$ *Lygris populata*	E	SA	10	0	-10	1.5	0.0	-1.5
$^{--}$ *Euphia molluginata*	S		9	3	-6	1.5	0.3	-1.2
$^{-}$ *Coenothephria verberata*	E		7	3	-4	1.3	0.3	-1.0
$^{-}$ *Perizoma blandiata*	E	SA	6	2	-4	1.3	0.2	-1.1
Lasiocampidae								
$^{-}$ *Malacosoma alpicola*	E	SA	8	3	-5	1.2	0.3	-0.9
Noctuidae								
$^{-}$ *Euclydia glyphica*	E		11	11	0	3.4	2.2	-1.2
$^{-}$ *Autographa gamma*	E		11	10	-1	3.1	2.2	-0.9
$^{-}$ *Chersotis cuprea*	S	SA	10	5	-5	1.7	0.5	-1.2
$^{-}$ *Phytometria viridaria*	E		6	4	-2	1.0	0.4	-0.6
Zygaenidae								
$^{-}$ *Zygaena purpuralis*	S		8	8	0	2.5	1.5	-0.9
$^{--}$ *Zygaena exulans*	E	SA	9	1	-8	2.2	0.1	-2.1
$^{-}$ *Zygaena filipendulae*	E		8	6	-2	2.0	0.7	-1.3
$^{-}$ *Zygaena loti*	S		7	4	-3	1.9	0.6	-1.3
$^{--}$ *Adscita alpina/geryon*	S		10	6	-4	2.6	0.9	-1.7

This means that the lower and upper range margins of particular species shifted upwards between 1977-79 and 2002-04. In 2002-04, for example, the community of 1977-79 in site M5, situated at 1760 m a.s.l., shows similarities with the community of 1977-79 in the sites F2 (situated at 1560 m a.s.l.) and P1 (situated at the north facing slope at 1340 m a.s.l.). This indicates that species which occurred at lower elevations
in 1977-79, have ascended into higher regions over the last 25 years. At the same time, species, which occurred at high altitudes in 1977-79, experienced losses. At the lower border of the study area, e.g. in site P2 and P3 (1480 m and 1490 m), the communities found in 2002-04 had increased their dissimilarity with the communities of 1977-79 at high elevations, such as M5 and M4. This indicates that species reaching their upper border of distribution in 1977-79 at these elevations, have increased and new species have immigrated into these habitats. This immigration can only be taking place from regions at lower altitude than that of the study area.

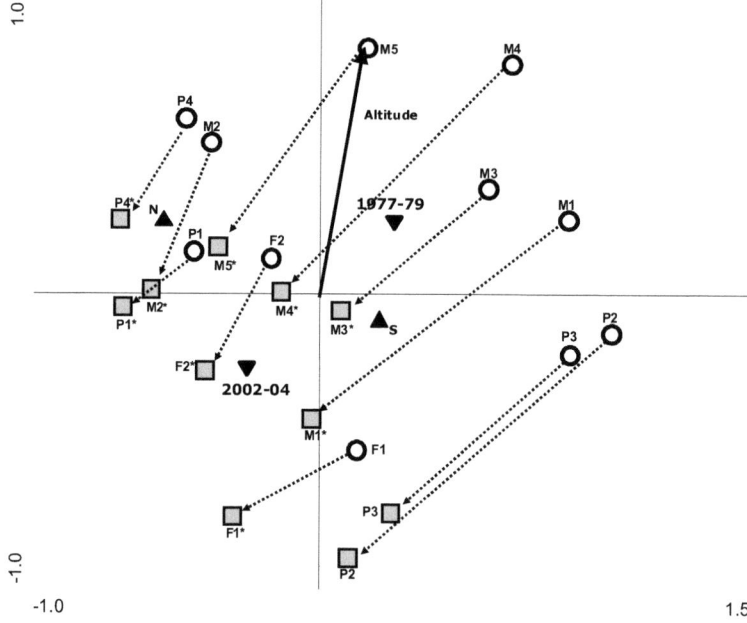

Figure 1. Changes in the eleven butterfly communities from 1977-79 (○) to 2002-04 (□). ▼ = centroids of the variable time. Passively added to the ordination diagram of the PCA are exposure (**N** = North facing slope; **S** = south facing slope) and altitude. Dotted arrows connect corresponding communities of 1977-79 and 2002-04 and represent the change over time (* = management in the site changed between 1977-79 and 2002-04).

Figure 2 is based on the same PCA analysis as Figure 1, but focuses on individual species. Shown are species increasing or decreasing from 1977-79 to 2002-04 in a statistically significant way according to the Wilcoxon test (Table 4).

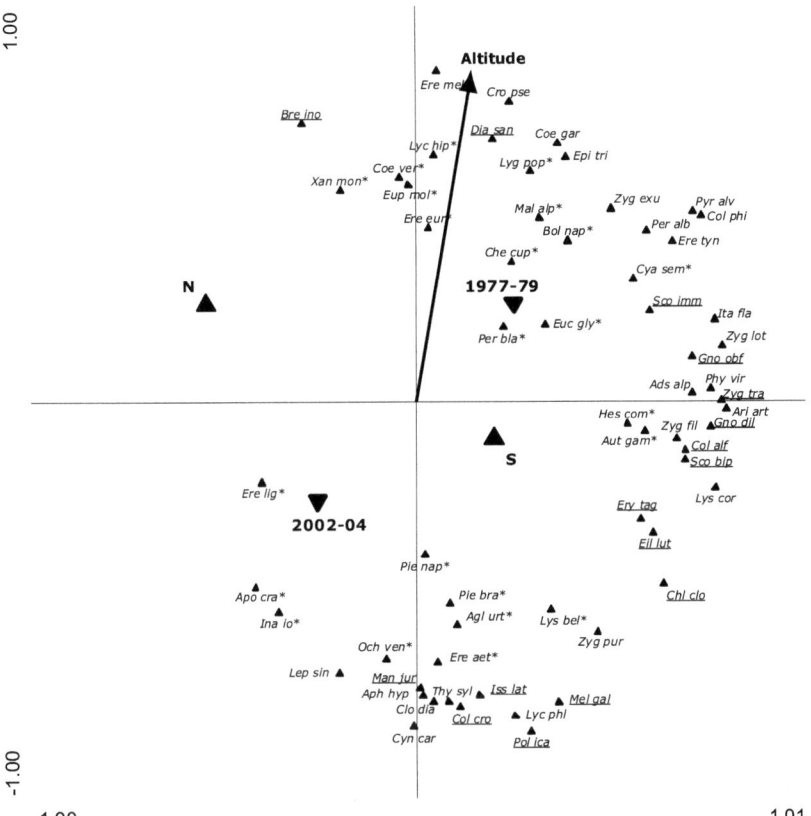

Figure 2. Ordination diagram of the PCA showing the butterfly species distribution across space and time. According to the Wilcoxon test, species with * significantly decreased and increased between 1977-79 and 2002-04 (p < 0.05). Underlined are species having not change significantly, but showing an important spatial variation between 1977-79 and 2002-04, i.e. they only showed clear changes in a few particular sites (used criterion for this categorisation = percentage fit of species in the ordination ≥ 50%). Species without a sign are fulfilling both criteria a significant change according to the Wilcoxon test and an important spatial variation between 1977-79 and 2002-04. For abbreviation of the species names see table 4 and 5. For other abbreviations see Fig.1.

The decreasing species are displayed mainly in the upper right part of the ordination, increasing species at the bottom (Fig. 2). Mainly species formerly abundant in the sites on the south facing slope experienced losses, e.g. *Colias phicomone*, *Coenonympha gardetta* and *Epirrhoe tristata*, particularly in the extensively managed meadows M1-M5. In the pastures at lower altitudes (P2, P3) species such as *Polyommatus coridon*, *Zygaena filipendulae* and *Adscita alpina/geryon* decreased.

The partial Redundancy analyses showed that time (1977-79 and 2002-04) and space (plot) together account for 76.6 % of the total variation in the species data. The factor time explains more than 15 % of the variation. The change of species composition between 1977-79 and 2002-04 is statistically significant.

The composition of the butterfly communities on the north-facing slope is significantly different from that on the south-facing slope. Exposure explains more than 16 % and altitude 10 % of variation. However, the species composition does not change in a statistically significant way along the altitudinal gradient.

In 2002-04, increasing species, like *Clossiana dia*, *Aphantopus hyperantus* and *Lycaena phlaeas* became particularly abundant in the pastures and in the formerly fertilised meadow (F1) at lower altitudes.

Additionally underlined in Figure 2 are species that have not changed significantly in abundance over time according to the Wilcoxon test, but show high percentage of fit with the variable time ($\geq$ 50 %) in the PCA (Table 5). This indicates that the species show drastic changes in only a few particular sites.

Table 5. List of species showing no significant increases or declines from 1977-79 to 2002-04 according to the Wilcoxon test, but high percentage fit of variance with the variable time (> 50 %) in the Principal Component Analysis (Fig.1 and 2). This indicates that these species show a pronounced temporal change in a few particular sites. Shown are the number of occupied sites and the mean class of abundance in 1977-79 and in 2002-04.

	Number of occupied sites			Mean class of abundance		
	1977-79	2002-04	Change	1977-79	2002-04	Change
Rhopalocera & Hesperiidae						
Colias alfacariensis/hyale	10	10	0	2.5	2.2	-0.3
Colias crocea	5	8	+3	0.6	1.1	0.5
Melanargia galathea	5	6	+1	1.2	1.6	0.5
Brenthis ino	8	9	+1	2.1	1.7	-0.4
Issoria lathonia	5	7	+2	0.5	1.0	0.5
Maniola jurtina	6	9	+3	1.4	2.2	0.8
Polyommatus icarus	8	9	+1	1.7	2.0	0.3
Erynnis tages	5	8	+3	0.9	1.1	0.2
Geometridae						
Scopula immorata	10	9	-1	1.9	1.0	-0.9
Scopula bipunctaria	5	3	-2	1.2	0.3	-0.9
Gnophos obfuscatus	7	3	-4	1.2	0.3	-0.9
Gnophos dilucedaria	5	4	-1	1.1	0.4	-0.7
Chlorissa chloraria	5	4	-1	0.7	0.5	-0.2
Arctiidae						
Eilema lutarella	4	5	+1	0.7	0.6	-0.1
Diacrisio sannio	7	4	-3	1.1	0.4	-0.7
Zygaenidae						
Zygaena transalpina	7	7	+0	2	1.2	-0.8

Indicated by the position of these species in Figure 2, such pronounced spatial trends of temporal changes from 1977-79 to 2002-04 were observed particularly in the lower sites of the valley, i.e. in the two pastures and in the fertilised meadow. The three Geometridae *Gnophos obfuscatus, G. dilucidaria, Scopula bipunctaria* and the Pieridae *Colias alfacariensis/hyale* experienced heavy losses here. On the other hand, other species became particularly abundant in these sites in 2002-04, e.g. *Melanargia galathea, Maniola jurtina* and *Polyommatus icarus*.

Temporal changes of distribution area and the abundance of stenotopic and eurytopic species

From the 46 species showing a statistically significant temporal change in abundance (Wilcoxon test, Table 4), 18 species were classified as stenotopic and 28 as eurytopic species.
Among the species showing changes from 1977-79 to 2002-04, significantly more stenotopic grassland species showed decreases than eurytopic species (Fisher exact test, $p< 0.05$).
According to the Wilcoxon test, species stenotopic on extensively managed grassland, such as *Pyrgus alveus, Epirrhoe tristata* or *Adscita alpine/geryon* occupied significantly fewer sites in 2002-04 than in 1977-79 (Fig. 3A, Table 4). On the other hand, eurytopic species, like *Leptidea sinapis, Thymelicus sylvestris* and *Erebia aethiops* tended to occupy more sites in 2002-04 than 1977-79.
The analysis of the mean class of abundance of stenotopic and eurytopic species showed a different pattern (Fig. 3B). The average mean class of abundance of stenotopic species significantly decreased from 1977-79 to 2002-04. Eurytopic species showed no significant changes in the mean class of abundance over the same time period, but their abundance tended to decrease.

Temporal changes in distribution area and abundance of colline-montane and subalpine-alpine species

Out of 46 species showing significant changes, only 23 species could be clearly classified as either colline-montane or subalpine-alpine (Table 4). The other species could not be assigned to these two groups. Without any exception, the species from the colline-montane zone increased and the 15 subalpine-alpine species have all declined between 1977-79 and 2002-04.
The analyses showed that colline-montane species, such as *Clossiana dia, Inachis io* and *Polyommatus bellargus* occured in significantly more sites in 2002-04 than in 1977-79, and increased in abundance (Fig. 4A, B). On the other hand, subalpine-alpine species like *Colias phicomone, Malacosoma alpicola* and *Zygaena exulans*, occupied significantly fewer sites in 2002-04 than in 1977-79. At the same time, they decreased significantly in the mean class of abundance,

as can also be observed for *Coenonympha gardetta*, *Erebia melampus* and *Erebia euryale* (see Table 4).

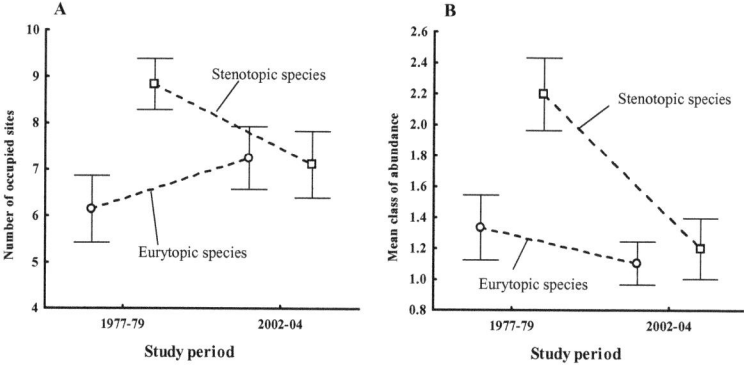

Figure 3. Species response from 1977-79 to 2002-04 according to grassland specialisation. (A) Change of average number of occupied sites by stenotopic grassland species and eurytopic species and (B) change of mean class of abundance. Δ number of occupied sites: stenotopic species, $p < 0.05$; eurytopic species, $p =$ ns. Δ mean class of abundance: stenotopic species, $p < 0.05$; eurytopic species = ns. Bars denote standard errors

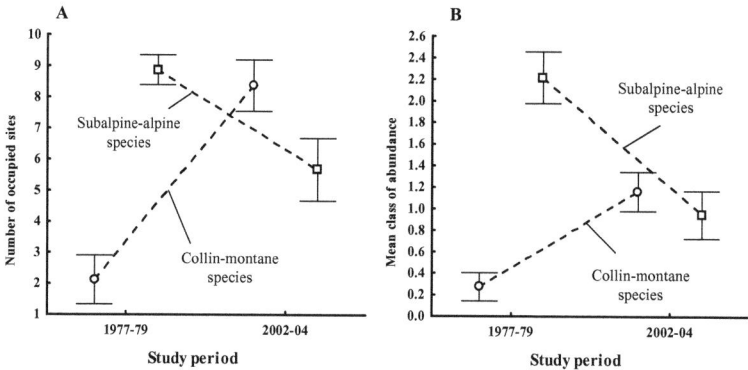

Figure 4. Species response from 1977-79 to 2002-04 according to their altitudinal preference. (A) Change of average number of occupied plots by colline-montane and subalpine-alpine species and (B) change of mean class of abundance. Δ number of occupied sites: colline-montane species, $p < 0.05$; subalpine-alpine species, $p < 0.05$. Δ mean class of abundance: colline-montane species, $p < 0.05$; subalpine-alpine species $p < 0.05$. Bars denote standard errors.

Discussion

Changes in Lepidoptera communities in Tavetsch

Over the last decades, butterfly monitoring programs in European countries and regions have shown an overall decline of butterfly populations (Hambler & Speight 1995; Asher *et al.* 2001; Maes & Van Dyck 2001; Warren *et al.* 2001; Conrad *et al.* 2004; Groenenendijk & Van der Meulen 2004). In the Alps, declines of butterflies and diurnal moths have, to our knowledge, for the fist time been detected in the present study. Out of 137 species, recorded in cultivated grassland in the Tavetsch valley in 1977-79, 15 species were no longer found in the 2002-04 census. On the other hand, 11 species not recorded in 1977-79 seem to have colonised the area over the last 25 years. 46 species experienced a statistically significant change in abundance. Two thirds of these changes were declines; butterflies showed declines as well as increases, diurnal moths only declines.

The loss and fragmentation of natural and semi-natural habitats as well as climate change are the most important causes for the recent changes in butterfly distribution and abundance in Europe (Pollard & Yates 1993; Pollard *et al.* 1995; Hill *et. al.* 2001a, b and 2002; Parmesan 2001, Warren *et al.* 2001; Konvicka *et al.* 2003). The observed changes in Lepidoptera communities in semi-natural grasslands in the subalpine zone of the Alps indicate both habitat degradation and effects of global warming.

Effects of the management

The decrease of the stenotopic species restricted to extensively managed grasslands indicates that the change of land use, i.e. intensification and abandonment took place in the valley over the last 25 years. This was already expected by Erhardt (1985a, b). The sharp declines of these species' abundances are most likely caused by a generally intensified grassland use since 1977-79.

Some of the observed declines can be linked to the change from mowing to grazing. In 1977-79, seven of the investigated sites were hay-meadows. Since then, five of them were grazed. Fewer individuals of butterflies are found in intensively managed pastures than in unfertilised hay-meadows, because such pastures support lower abundances of host and nectar plants than unfertilised hay-meadows (Loertscher *et al.* 1995; Wettstein & Schmid 1999). However, the number of butterfly species occurring in lightly managed pastures, can be similar or even higher than in unfertilised hay-meadows (Erhardt & Thomas 1991; Erhardt 1995; Saarinen & Jantunen 2005). This might be a reason why the average number of species found per site has not decreased over the last 25 years.

Some stenotopic grassland species may even fare better under the new rather than the former management conditions. A good example of such a species is *Polyommatus bellargus* (= *Lysandra bellargus*). The increase of this species might be related to an increased number of grazed grasslands. At the hatching time in the area (late June) the females find large areas of grazed turfs. These are the preferred places for ovi-position of this species (Thomas 1983).

Large changes of species composition were also observed in the three unfertilised hay meadows of 1977-79, which were still mown in 2002-04. Compared to 1977-79 when they were mostly mown every second year, these meadows have been mown annually for about ten years. Erhardt (1985a, b) showed that Lepidoptera benefit if the vegetation is not disturbed, in this case not mown, for one or two years. In such grasslands, host plants and nectar resources remain undisturbed. Thus, we assume that the annual mowing of these meadows may at least partially account for the observed declines of stenotopic grassland species.

In all the investigated grasslands, woody plants (e.g. *Picea abies, Betula pendula, Alnus viridis, Rosa canina*) typically occurring in fallow grassland at these altitudes, slightly increased from 1977-79 to 2002-04, even in annually mown meadows.

Many butterfly species benefit from an increasing shrub cover in semi-natural grasslands, at least up to a critical value of cover (Erhardt 1985a, b; Balmer & Erhardt 2000; Söderström *et al.* 2001). In the two investigated pastures on the north-facing slope the increase of shrubs was clearly caused by abandonment. In grasslands where the management has changed from mowing to grazing, the increase of woody plants can be explained by the patchy and selective grazing of cattle and sheep. In pastures, woody plants increase if the farmers do not remove them from time to time because cattle and sheep do not eat them. As mowing prevents the growth of woody plants these meadows were most likely abandoned for a few years between 1977-79 and 2002-04, although this was not always clearly confirmed by the owners of the meadows.

This increase of shrubs and trees is, however, unlikely to have caused the observed declines in grassland species, as such plants did not cover major parts of the investigated grasslands. However, over a long-term perspective these semi-natural habitats will lose their grassland character and its typical Lepidoptera communities (Erhardt 1985a, b; Balmer & Erhardt 2000) if woody plants should further increase in the sites.

Effects at landscape scale

Although no direct impact of woody plants in the investigated grasslands could be demonstrated, the general increase of shrubs and forests in the Tavetsch certainly affected its Lepidoptera communities. Already in 1977-79, large areas of formerly extensively managed grasslands had been abandoned (Erhardt 1985b). Since then, the abandonment has progressed, and the early stages of abandonment have turned into later successional stages and climax vegetation i.e. forests at the expense of grassland habitats. This process was further amplified by afforestations of grasslands for protecting villages and traffic infrastructure against avalanches.

It is well known that the surrounding landscape can influence the composition of insect communities observed at a particular site (Jeanneret *et al.* 2003a, b). The majority of the investigated meadows and pastures of 1977-79 is still cultivated in 2002-04. Hence, the observed increase of species preferring woodland borders and hedgerows must be linked mainly to the increase of shrubs and forests in the surrounding of the investigated meadows and pastures. Good examples of such species are *Erebia ligea* and *Leptidea sinapis* (*Lepidopterologen-Arbeitsgruppe* 1991).

On the other hand, it is very probable that the increase of woodlands has increased the isolation of extensively managed meadows and pastures and thus additionally lowered the occurrence of specialised grassland species (see also Wettstein & Schmid 1999).

Effects of climate warming

Parallel to the described habitat deterioration for Lepidoptera, the observed changes in subalpine Lepidoptera communities suggest effects of climatic warming. Overall, we observed an upward shift of butterfly species in the Tavetsch valley. Typical colline-montane species increased in abundance and occurred at higher elevations in 2002-04 than in 1977-79. At the same time, species restricted to the cooler habitats in the subalpine-alpine zone experienced losses.

Over the last decades, the analysis of historical and recent distribution maps across wide latitudinal ranges showed a northward expansion of butterflies and other taxa that can be explained by global warming (Parmesan *et al.* 1999; Thomas & Lennon 1999, Hill *et al.* 2001a; Hill *et al.* 2002; Parmesan & Yohe, 2003, Root *et al.* 2003). Studies on altitudinal shifts of species are scarce, however, because effects of a climatic warming on the altitudinal distribution of species concern relatively short distances. Thus, reliable comparisons require historical data of high quality and on fine scales (Konvicka *et al.* 2003). Hill *et al.* (2002) found altitudinal shifts of butterflies in Britain. Konvicka *et al.* (2003) showed that particular butterflies colonised higher elevations during the last half of the 20th century in the Czech Republic. Our study confirms the uphill shifts of Lepidoptera also in a European alpine area.

In the summer of 2003, temperatures were exceptionally high on average in Switzerland and Europe in general (ProClim- 2005). One may ask whether the overall changes found in the Lepidoptera communities from 1977-79 to 2002-04 were influenced by this hot summer. Thomas *et al.* (2001)

showed that changes of environmental conditions at existing range margins can initialise increased dispersal behaviour of butterflies. Complementary analyses of our data on the species grouped according to their altitudinal preferences have shown that the year 2003 differed not significantly from neither 2002 nor 2004 in the number of occupied sites and the mean class of abundance. Furthermore, overall upwards shifts could already be observed comparing the surveys of 1977-79 with each individual year from 2002 to 2004. Therefore, we assume that the hot summer of 2003 has not biased the generally observed pattern.

Implications of the combined effects of land use change and climate warming

Our study in the Tavetsch valley indicates that in the subalpine zone, land use change and climate warming alter the abundance and distribution of Lepidoptera species. Spehn *et al.* (2002) postulate that land use change is the most important factor of all global change impacts on mountain biodiversity. The changes in the Lepidoptera communities of the Tavetsch valley lead to a different conclusion. The increase of colline-montane species and the decrease of subalpine-alpine species are more pronounced than the changes in stenotopic and eurytopic species regarding management types. This suggests that the land use change in the valley had a less severe impact on the changes of the Lepidoptera communities than the climatic changes of the last decades.

Our findings in the Tavetsch valley show similarities to those of Warren *et al.* (2001) in Britain, where several European butterflies reach their northern range margin. At this margin, it can be expected that butterflies benefit from a climate warming (Warren *et al.* 2001). Warren *et al.* (2001) showed that the combined impacts of habitat degradation and climatic warming at the northern range margin of the European butterfly fauna is likely to cause specialised species to decline, leaving biological communities dominated by mobile and widespread habitat generalists. Our results show a similar pattern. The majority of the lowland species showing increases are eurytopic; according to Gonseth (1987) they are widespread in the Swiss lowlands. On the other hand, species restricted to extensively managed grassland experienced losses. This pattern is likely caused by the fact that specialised species are often sedentary and fail to expand in fragmented landscapes due to the isolation of their habitats, whereas generalists are relatively mobile and disperse rapidly into new areas (Thomas *et al.* 2001; Warren *et al.* 2001).

Nature Conservation

The high number of species found in the meadows and pastures of our subalpine valley in 2002-04 shows that these habitats are still of high conservation value. 122 of 137 species found in 1977-79 were still present in 2002-04, including species, which were rated rare in 1977-79. Out of 77 species of Rhopalocera & Hesperiidae found in 2002-04, 26 species are on the Red List of endangered butterflies in Switzerland (Gonseth 1994) and five species are listed in the categories 1-3 in the Red Data Book of Europe (Van Swaay & Warren 1999). Two species, *Maculinea arion* and *Parnassius apollo* are mentioned in the Appendix II of the Bern Convention. However, over the last 25 years, changes occurred in subalpine Lepidoptera communities and their habitats in the investigated valley and these are causing concern. Most experts agree that 'butterflies are reasonable representatives for other insects' (Thomas 2005). Thus, it must be assumed that other insects in semi-natural grasslands of the subalpine zone have reacted or will react similarly to the observed land use and climatic changes.

In 2002-04, most of the extensively managed grasslands in the valley were part of the Swiss agri-environmental programme, which was introduced in 1992-93 in order to maintain traditional cultivation methods and species rich grasslands in Switzerland (Bundesblatt 1992). Due to compensation payments for farmers, this programme has partially ensured the agricultural management of sites difficult to manage and hence prevented many grassland habitats from abandonment over the last decade (Kampmann & Herzog 2006.). However, the observed declines of diurnal Lepidoptera suggest that the agricultural programme favours rather intensive grazing instead of light grazing or mowing without fertilising in mountain regions. Some mandatory instructions for managing grasslands given by the programme must be assessed as inefficient in preserving Lepidoptera communities or even threatening to specialised species. Erhardt (1985b) showed in detail the importance of a short-time abandonment of these meadows for Lepidoptera. However, these meadows are mown annually at present because farmers are obligated to manage them in order to receive compensation payments.

Our study shows that there is an urgent need for butterfly conservation plans and Lepidoptera friendly management advice for extensively managed grassland in the subalpine zone of the Swiss Alps. The habitat requirements of these insects should be taking into account in existing agri-environmental programmes and conservation plans. In the subalpine zone, unfertilised hay-meadows should only be mown every second year. A rotational management plan for existing pastures and abandoned grassland should be considered including the management of shrub and tree encroachment (see also Balmer & Erhardt 2000). In order to increase the ability of species to colonise higher elevated regions, large-scale networks of extensively managed grasslands across the different altitudinal zones (colline-montane-subalpine-alpine) are required (see also Warren *et al.* 2001).

Acknowledgements

We are grateful to the farmers in the Tavetsch valley for allowing us to work on their land. We thank Eugen Pleisch, Ladislaus Reser and Andreas Müller for help with identifying Lepidoptera, Jessica Beller, Markus Peter and Dorothea Kampmann for helping in the field and Catherine Palmer the language corrections. This study was financed by Swiss National Science Foundation (grant 4048-064405).

Parnassius apollo L.

References

Asher, J., Warren, M. S., Fox, R., Harding, P., Jeffcoate, G. and Jeffcoate, S. 2001. The millennium atlas of butterflies in Britain and Ireland. Oxford University Press,   Oxford, New York.

Balmer, O. and Erhardt, A. 2000. Consequences of succession on extensively grazed   grasslands for central European butterfly communities: Rethinking conservation   practices. Conservation Biology 14: 746-757.

Beniston, M. 2003. Climatic change in mountain regions: A review of possible impacts. Climate Change 59: 5-31.

Bischof, N. 1981. Gemähte Magerrasen in der subalpinen Stufe der Zentralalpen. Bauhinia 7   (2): 81-128.

Bischof, N. 1984. Pflanzensoziologische Untersuchungen von Sukzessionen aus gemähten Magerrasen in der subalpinen Stufe der Zentralalpen. Beiträge zur geobotanischen Landesaufnahme der Schweiz 60: 1-128.

Bundesblatt 1992. Botschaft zur Änderung des Landwirtschaftgesetzes vom 27. Januar 1992. Bundeskanzlei, BBL II (92.010): 1-132.

Burga, C. A., Haeberli, W., Krummenacher, B. and Walther, G.-R. 2003. Abiotische und biotische Dynamik in Gebirgsräumen – Status quo und Zukunftsperspektiven. Pages 25-37 in Jeanneret, F., Wastl-Walter, D., Wiesmann, U. and Schwyn, M., editors. Welt der Alpen – Gebirge der Welt. Ressourcen, Akteure, Perspektiven. Haupt, Bern.

Conrad, K. F., Woiwod, I. P., Parsons, M., Fox, R. and Warren, M. S. 2004. Long-term population trends in widespread British moths. Journal of Insect Conservation 8: 119- 136.

Ebert G. 1991-2003. Die Schmetterlinge Baden-Württembergs. Band 1-9, Ulmer, Stuttgart.

Ehrlich, P. R. 1994. Energy use and biodiversity loss. Philosophical Transactions of the Royal Society, Biological Science 344: 99-104.

Erhardt, A. 1985a. Diurnal Lepidoptera: Sensitive indicators of cultivated and abandoned grassland. Journal of Applied Ecology 22: 849-861.

Erhardt, A. 1985b. Wiesen und Brachland als Lebensraum für Schmetterlinge. Birkhäuser, Basel.

Erhardt, A. and Thomas J. A. 1991. Lepidoptera as indicators of change in the semi-natural grasslands of lowland and upland of Europe. Pages 213-236 in Collins, N. M. and Thomas, J. A., editors. The Conservation of insects and their habitats, Academic Press, London.

Erhardt, A. 1995. Ecology and conservation of alpine Lepidoptera. Pages 258-276 in Pullin A. S., editor. Ecology and conservation of butterflies. Chapman & Hall, London.

Forster, W. and Wohlfahrt, T. A. 1954-81. Die Schmetterlinge Mitteleuropas. Band 1-5, Franckh'sche Verlagshandlung Stuttgart.

Gonseth, Y. 1987. Verbreitungsatlas der Tagfalter der Schweiz (Lepidoptera, Rhopalocera). Documenta Faunistica Helvetiae, CSCF, Neuchâtel, 6: 1-242.

Gonseth, Y. 1994. Rote Listen der gefährdeten Tagfalter der Schweiz. Pages 48-51 in P. Duelli, editor. Rote Listen der gefährdeten Tierarten der Schweiz. Bundesamt für Umwelt, Wald und Landschaft, Bern, Switzerland.

Grabherr, G., Gottfried, M. and Pauli, H. 1994. Climate effects on mountain plants. Nature 369: 448-449.

Groenendijk, D. and Van der Meulen, J. 2004. Conservation of moths in The Netherlands: population trends, distribution patterns and monitoring techniques of day-flying moths. Journal of Insect Conservation 8: 109-118.

Guisan, A., Holten, J. I., Spichiger, R. and Tessier, L. 1995. Potential ecological impacts of climate change in the Alps and Fennoscandian Mountains. Editions des Conservatoire et Jardin botaniques, Genève.

Hambler C. and Speight M. R. 1996. Extinction rates in British nonmarine invertebrates since 1900. Conservation Biology 10: 892-896.

Hartmann, J. 1976. Mähwiesen und Sozialbrachen im Tavetsch. Diplomarbeit, University of Basel, Basel.

Hill, J. K., Thomas, C. D. and Huntley, B. 2001a. Climate and recent range changes in butterflies. Pages 77-88 in Walther, G.-R., Burga, C. A. and Edwards, P. J., editors. Fingerprints of climate change – Adapted behaviour and shifting species ranges. Kluwer/Plenum, New York and London.

Hill, K. H., Collingham, Y. C., Thomas, C. D., Blakeley, D. S., Fox, R., Moss, D. and Huntley, B. 2001b. Impacts of landscape structure on butterfly range expansion. Ecology Letters 4: 313-321.

Hill, J. K., Thomas, C. D., Fox, R., Telfer, G., Willis, S. G., Asher, J. and Huntley B. 2002. Responses of butterflies to the twentieth century climate warming: implications for future ranges. Proceedings of the Royal Society of London, Series B-Biological Sciences 269: 2163-2171.

Jeanneret, P., Schüpbach, B., Pfiffner, L., Herzog, F. and Walter T. 2003a. The Swiss agri-environmental programme and its effects on selected biodiversity indicators. Journal for Nature Conservation 11: 213-220.

Jeanneret, P., Schüpbach, B., Pfiffner, L. and Walter T. 2003b. Arthropod reaction to landscape and habitat features in agricultural landscapes. Landscape Ecology 18: 253- 263.

Kampmann, D. and Herzog F. 2006. Ökomassnahmen im Bergebiet erhalten die Artenvielfalt. Agrarforschung 13; 56-61.

Konvicka, M., Maradova, M., Benes, J., Fric, Z. and Kepka, P. 2003. Uphill shifts in distribution of butterflies in the Czech Republic: effects of changing climate detected on a regional scale. Global Ecology and Biogeography 12: 403-410.

Lauber, K. and Wagner, G. 2001. Flora Helvetica. Haupt, Bern, Switzerland.

Lepidopterologen-Arbeitsgruppe 1991-2000. Tagfalter und ihre Lebensräume. Vol. 1-3, Schweizerischer Bund für Naturschutz, Basel.

Leraut, P.J.A. 1997. Liste systématique et synonymique des Lépidoptères de France, Belgique et Corse. Suppl. à Alexanor, Paris-Wetteren.

Loertscher, M., Erhardt, A. and Zettel, J. 1995. Microdistribution of butterflies in a mosaic-like habitat: the role of nectar resources. Ecography 18: 15-26.

Maes, D. and Van Dyck, H. 2001. Butterfly diversity loss in Flanders (north Belgium): Europe's worst case scenario? Biological Conservation 99: 263-276.

Master, L. L., Stein, B. A., Kutner, L. S. and Hammerson, G. A. 2000. Vanishing assets. Pages 93-118 in Stein, B. A., Kutner, L. S. and Adams, J. S., editors. Precious heritage: the status of biodiversity in the United States, Oxford University Press, Oxford, United Kingdom.

Parmesan, C., Ryrholm, N., Stefanescu, C., Hill, J. K., Thomas, C. D., Descimon, H., Huntley, B., Kaila, L., Kullberg, J., Tammaru, T., Tennent, W. J., Thomas, J. A. and Warren, M. 1999. Poleward shifts in geographical ranges of butterfly species associated with regional warming. Nature 399: 579-583.

Parmesan, C. 2001. Detection of range shifts: General methodological issues and case studies of butterflies. Pages 57-76 in Walther, G.-R., Burga, C. A. and Edwards, P.J., editors. Fingerprints of climate change – Adapted behaviour and shifting species ranges. Kluwer/Plenum, New York and London.

Parmesan, C. and Yohe, G. 2003. A globally coherent fingerprint of climate change impacts across natural systems. Nature 421: 37-42.

Pollard, E. and Yates, T. J. 1993. Monitoring butterflies for ecology and conservation. Chapman & Hall, London.

Pollard, E., Moss, D. and Yates, T. J. 1995. Population trends of common British butterflies at monitored sites. Journal of Applied Ecology 32: 9-16.

ProClim- Forum for Climate and Global Change 2005. Hitzesommer 2003-Synthesebericht. Platform for the Swiss Academy of Sciences, Bern.

Root, T. L., Price, J. T., Hall, K. R., Schneider, S. H., Rosenzweig, C. and Pounds, J. A. 2003. Fingerprints of global warming on wild animals and plants. Nature 421: 57-60.

Saarinen, K. and Jantunen, J. 2005. Grassland butterfly fauna under traditional animal husbandry: contrasts in diversity in mown meadows and grazed pastures. Biodiversity and Conservation 14: 3201-3213.

Sokal, R. R. and Rohlf, F.J. 1981. Biometry. Freeman, San Francisco, California.

Söderström, B., Svensson, B., Vessby, K. and Glimskär, A. 2001. Plants, insects and birds in semi-natural pastures in relation to local habitat and landscape factors. Biodiversity and Conservation 10: 1839-1863.

Spehn, E., Messerli, B. and Körner, C. 2002. A global assessment of mountain biodiversity: synthesis. Pages 325-330 in Körner, C. and Spehn, E., editors. Mountain biodiversity: a global assessment. Parthenon, New York, London.

Ter Braak, C. J. F. and Prentice, I. C. 1988. A theory of gradient analysis. Advances in Ecological Research 18: 271-317.

Ter Braak, C. J. F. and Wiertz, J. 1994. On the statistical analysis of vegetation change: a wetland affected by water extraction and soil acidification. Journal of Vegetation Science 5: 361-372.

Ter Braak, C.J.F. and Šmilauer, P. (2002) CANOCO Reference Manual and CanoDraw for Windows - User's Guide: Software for Canonical Community Ordination, Version 4.5. Biometris, Wageningen and České Budejovice.

Theurillat, J. P. and Guisan, A. 2001. Potential impacts of climate change on vegetation in the European Alps: A review. Climate Change 50: 77-109.

Thomas, C. D. and Lennon, J. J. 1999. Birds extend their range margins. Nature 399: 213.

Thomas, C. D., Bodsworth, E. J., Wilson, R. J., Simmons, A. D., Davies, Z. G., Musche, M. and Conradt, L. 2001. Ecological and evolutionary processes at expanding range margins. Nature 411: 577-581.

Thomas, J. A. 1983. The ecology and conservation of *Lysandra bellargus* (Lepidoptera: Lycaenidae) in Britain. Journal of Applied Ecology 20: 59-83

Thomas, J. A. 2005. Monitoring change in the abundance and distribution of insects using butterflies and other indicator groups. Philosophical Transactions of the Royal Society, Biological Science 360: 339-357.

Thomas, J. A., Telfer, M. G., Roy, D. B., Preston, C. D., Greenwood, J. J. D., Asher, J., Fox, R., Clarke, R. T. and Lawton, J. H. 2004. Comparative losses of British butterflies, birds, and plants and the Global Extinction Crisis. Science 303: 1879-1881.

Tolman, T and Lewington, R. 1998. Die Tagfalter Europas und Nordwestafrikas, Kosmos, Stuttgart.

Van Swaay, C. A. M. and Warren, M. S. 1999. Red Data book of European butterflies (Rhopalocera). Nature and Environment, No. 99, Council of Europe Publishing, Strasbourg.

Walther, G.-R., Beissner, S. and Burga, C. A. 2005. Trends in the upward shift of alpine plants. Journal of Vegetation Science 16: 541-548.

Warren, M. S., Hill, J. K., Thomas, J. A., Asher, J., Fox, R. Huntley, B., Roy, D. B., Telfer, M. G., Jeffcoate, S., Harding, P., Jeffcoate, G., Willis, S. G., Greatorex-Davies, J. N., Moss, D. and Thomas, C. D. 2001. Rapid responses of British butterflies to opposing forces of climate and habitat change. Nature 414: 65-69.

Wettstein, W. and Schmid, B. 1999. Conservation of arthropod diversity in montane wetlands: effects of altitude, habitat quality and habitat fragmentation on butterflies and grasshoppers. Journal of Applied Ecology 36: 363-373.

Appendix

Appendix 1. List of species showing an increase from 1977-79 to 2002-04 and probable causes for this increase. Species are categorized in: Increase, $p \leq 0.05$; marginal increase, $0.05 > p \leq 0.1$; weak trends to increase, $p > 0.1$, according to the Wilcoxon test. Bold species names: species in the Red List categories 1-3 for the whole of Switzerland (in Gonseth 1994). Blank = cause unknown.

	Increasing species	Causes for the increases				
		Mowing to grazing	Biennial to annual mowing	Early stages of abandonement	Increasing of woodland in the surrounding	Climate warming
Increase						
Rhopalocera & Hesperiidae						
	Aglais urticae L.	x		x		
	Aphantopus hyperantus L.			x		x
	***Aporia crataegi* L.**			x	x	x
	***Clossiana dia* L.**			x	x	x
	Cynthia cardui L.			x		x
	***Erebia aethiops* ESP.**			x	x	x
	Erebia ligea L.				x	
	Inachis io L.	x				x
	Leptidea sinapis L.			x	x	x
	Lycaena phlaeas L.	x				x
	Polyommatus bellargus ROTT.	x				x
	Ochlodes venatus TURATI			x	x	x
	Pieris brassicae L.			x		
	Pieris napi L.			x		
	Thymelicus sylvestris PODA			x	x	
Marginal increase						
Rhopalocera & Hesperiidae						
	Clossiana euphrosyne L.				x	
	Heodes tityrus PODA	x				
	***Lasiommata petropolitana* F.**					
	Maniola jurtina L.					x
Geometridae						
	Cabera pusaria L.				x	
	Ematurga atomaria L.			x		
	Pseudopanthera macularia L.				x	x
	Scopula ornata SCOP.	x				x
Weak trends to increase						
Rhopalocera & Hesperiidae						
	Araschnia levana L.				x	
	Argynnis paphia L.				x	x
	***Callophrys rubi* L.**			x	x	
	***Clossiana selene* D.& S.**			x	x	
	***Clossiana titania* ESP.**			x	x	
	Colias crocea GEOFFROY					x
	Erynnis tages L.			x		
	***Eumedonia eumedon* ESP.**			x		
	***Heodes virgaureae* L.**			x		
	Euphydryas i.wolfensbergeri FREY.				x	
	Issoria lathonia L.					x
	Lasiommata maera L.	x				
	Lasiommata megera L.	x				x
	Melanargia galathea L.					x
	***Melitaea diamina* LANG**			x		
	***Mellicta athalia* ROTT.**			x		
	***Nymphalis antiopa* L.**				x	x
	Papilio machaon L.			x		
	Pararge aegeria L.				x	x
	***Parnassius apollo* L.**	x				
	Polyommatus eros OCHSENHEIMER					
	Polyommatus icarus ROTT.					x
	Pyrgus malvoides E.& E.					
	***Pyrgus serratulae* RAMBUR**					
	Vanessa atalanta L.	x				x

Appendix 1 continued

Increasing species	Causes for the increases				
	Mowing to grazing	Biennial to annual mowing	Early stages of abandonement	Increasing of woodland in the surrounding	Climate warming
Geometridae					
Chiasmia clathrata L.					x
Eupithecia satyrata HBN.			x		
Idea serpentata HUFN.			x		
Itame brunneata THNBG.			x		
Noctuidae					
Callestige mi CL.			x		
Euxo recussa HBN.					
Hadena confusa HUFN.					
Panemeria tenebrata SCOP.					x

Appendix 2: List of species showing a decrease from 1977-79 to 2002-04 and probable causes for declines. Species are categorized in: Decrease, p ≤ 0.05; marginal decrease, 0.05 > p ≤ 0.1; weak trends to decrease, p > 0.01, according to the Wilcoxon test. Bold species names: species in the Red List categories 1-3 for the whole of Switzerland (in Gonseth 1994). Blank = cause unknown.

Decreasing species	Causes for the decreases				
	Mowing to grazing	Biennial to annual mowing	Early stages of abandonement	Increasing of woodland in the surrounding	Climate warming
Decrease					
Rhopalocera & Hesperiidae					
Aricia artaxerxes GEYER.			x	x	x
Boloria napae HOFFMSG. / *Boloria pales* D. & S.			x	x	x
Coenonympha gardetta PRUN.	x	x		x	x
Colias phicomone ESPR.	x	x		x	x
Cyaniris semiargus ROTT.	x	x	x	x	
Erebia euryale adyte HBN.					x
Erebia melampus FUESSL.		x		x	x
Erebia tyndarus ESP.				x	x
Hesperia comma L.		x		x	
Palaeochysophanus hippothoe eurydame ROTT.	x			x	x
Polyommatus coridon PODA	x	x		x	
Pyrgus alveus HBN.		x		x	
Geometridae					
Crocota pseudotinctaria LERAUT	x	x		x	x
Epirrhoe molluginata HBN.					
Epirrhoe tristata L.		x		x	
Eulithis populata HBN.					x
Idea flaveolaria HBN.	x			x	x
Perizoma albulata D. & S.	x			x	
Perizoma blandiata D.&S.					x
Perizoma verberata SCOP.					
Xanthorhoe montanata D.&S.					
Noctuidae					
Autographa gamma L.		x		x	
Chersotis cuprea D. & S.	x	x		x	x
Euclidia glyphica L.		x		x	
Phytometra viridaria CL.	x	x		x	
Lasiocampidae					
Malacosoma alpicola STAUDINGER		x			x
Zygaenidae					
Adscita alpina ALBERTI / *Adscita geryon* HBN.	x	x		x	
Zygaena exulans HOHENWARTH					x
Zygaena filipendulae L.		x		x	
Zygaena loti D.& S.	x	x		x	
Zygaena purpuralis BRUENN.	x	x		x	

Appendix 2 continued

Decreasing species	Mowing to grazing	Biennial to annual mowing	Early stages of abandonement	Increasing of woodland in the surrounding	Climate warming
Marginal decrease					
Rhopalocera & Hesperiidae					
Maculinea arion L.			x	x	
Pieris rapae L.					
Thymelicus lineola OCHSENH.	x			x	
Vacciniina optilete KNOCH					x
Geometridae					
Aplocera praeformata HBN.					
Entephria caesiata D.&S.					
Gnophos obfuscatus THNBG.					
Idea humiliata HFN.					
Odezia atrata L.	x			x	
Parietaria dilucidarius D. & S.					
Scopula immorata L.	x	x		x	
Scotopteryx bipunctata D.&S.	x	x		x	
Scotopteryx chenopodiata L.			x	x	
Xanthorhoe spadicearia D.&S.					
Noctuidae					
Cerapteryx graminis L.				x	
Arctiidae					
Eilema lurideola ZINCKEN					
Parasemia plantaginis L.					
Rhopalocera & Hesperiidae					
Anthocharis cardamines L.					x
Brenthis ino ROTT.			x	x	
Colias alfacariensis RIBBE / Colias hyale L.	x			x	
Colias palaeno ESPR.				x	x
Cupido minimus L.	x	x		x	
Erebia eriphyle FR.					x
Erebia manto D.& S.	x			x	x
Erebia montana PRUN.				x	x
Erebia pharte HBN.				x	x
Euphydryas aurinia debilis OBTH.		x		x	
Pieris bryoniae HBN.					x
Plebejus argus L.	x			x	
Plebejus idas L.				x	
Polyommatus dorylas D. & S.	x			x	
Pseudophilotes baton BERGSTRÄSSER				x	
Speyeria aglaia L. / Fabriciana niobe L.	x	x			
Geometridae					
Chlorissa cloraria HBN.				x	
Chloroclysta citrata L. / Chloroclysta truncata HFN.					
Colostygia aptata HBN.					
Eupithecia plumbeolata HAWORTH					
Psodos quadrifaria SULZER	x			x	
Scopula incanata L.					
Scopula ternata SCHRK.					
Noctuidae					
Autographa aemula D. &. S.					
Autographa bractea D. &. S.				x	
Chersotis ocellina D. &. S.					
Eriopygodes imbecilla F.					
Leucania comma L.					
Omia cymbalariae HBN.	x			x	
Lasiocampidae					
Lasiocampa quercus L.					x
Arctiidae					
Diacrisia sannio L.	x	x		x	
Eilema cereola HBN.					
Zygaenidae					
Zygaena lonicerae SCHEVEN	x	x		x	
Zygaena transalpina ESP.	x	x		x	

Appendix 3: Species showing no changes between 1977-79 and 2002-04. Bold species names: species in the Red List categories 1-3 in Gonseth (1994).

	Species
Rhopalocera & Hesperiidae	***Erebia oeme* HBN. / *Erebia meolans* PRN.**
Geometridae	*Perizoma minorata* TREISCHKE
Noctuidae	*Heliothis ononis* D. & S.
Arctiidae	*Eilema lutarella* L.

II. Comparison of grasshopper assemblages recorded in 1981-83 and in 2002-03 as an indication of land use change in Grindelwald, northern Swiss Alps

M. Hohl[1], P. Jeanneret[1], T. Walter[1], and A. Gigon[2]

[1]Agroscope FAL Reckenholz, Swiss Federal Research Station for Agroecology and Agriculture, Reckenholzstr. 191, CH-8046 Zurich, Switzerland
[2]Institute of Integrative Biology, Swiss Federal Institute of Technology, ETH Zurich, CH-8092 Zurich, Switzerland

Abstract

Changes in grassland management are expected to continue in the Swiss Alps. These grasslands are hotspots for biodiversity and represent important habitats for grasshoppers. The transformation of extensively managed grassland into forests following abandonment, or into intensively used meadows and pastures threaten this biodiversity. Taking into account the spatial variation of grasshopper occurrence, we compared grasshopper assemblages of differently managed grasslands recorded between 1981-83 with records made in 2002-03, using the same twenty plots in Grindelwald (northern part of the central Swiss Alps) and the same method. The plots were situated in three different altitudinal zones between 900 and 2000 m and they were of different steepness.

The spatial analysis showed that both altitude and slope have a significant effect on the composition of the species assemblages. The assemblages in the upper montane zone (1100-1500 m) are the most species rich with an average of 14 species per plot. In 2002-03, all 18 species found in 1981-83 were found again. The difference between the grasshopper assemblages of 1981-83 and 2002-03 is very small, but statistically significant. Some species no longer occur locally in particular plots in 2002-03, but are found in other plots in 2002-03 compared to 1981-83.

The effect of slope on the composition of grasshopper assemblages was almost as strong as the well-known effect of altitude. The grasshoppers profit from the extensive management in steep plots. In the upper montane zone a high number of species are found, because in this zone, lowland and subalpine species occur at the upper and lower border of their distribution and the grasslands here are managed extensively.

The comparison of the species composition over time indicated no substantial differences between the species composition of 1981-83 and 2002-03. This suggests that no general change of land use with detrimental effects on grasshoppers has occurred over the last two decades in the Grindelwald region. The high number of grasshopper species in the plots observed in 2002-03 rather indicates a very appropriate management from the grasshopper's perspective and we assume that this management has been practiced over the last two decades. Thus, the agricultural management types play an important role in maintaining grasshopper diversity in the Grindelwald region and should be taken into account when planning conservation activities. Special focus should be given to extensively used meadows and pastures in the upper montane zone.

Keywords: grassland management, agriculture, altitudinal zones, biodiversity, grasshopper conservation

Grindelwald Vally (Bernese Highlands, Switzerland)

Introduction

In the Swiss Alps, different types of grasslands cover more than 50% of the area and represent the major form of agricultural land use (BFS 1997). These habitats have evolved as a result of extensive agriculture over hundreds of years and they represent hotspots of biodiversity in Switzerland (Körner 1995; Theurillat et al. 1995). Due to ongoing economical and political changes in agriculture, changes in grassland management are expected to continue. Depending on the area's productivity, formerly extensively managed grassland is given up and either abandoned and consequently converted into forests, or is followed by an increased management (Gotsch et al. 2004; Garcia-Ruiz et al. 1996). From a long-term perspective both abandonment and agricultural intensification pose a threat to grassland biodiversity (Zoller & Bischof 1980; Erhardt 1985; Dietl 1995). Up to now, information has been scarce in regard to assessing the consequences of these developments for biodiversity over the last two decades in the Swiss Alps, in particular for fauna. A national biodiversity monitoring program, which also uses vertebrates and invertebrates as indicators for biodiversity, has been started quite recently (Hintermann et al. 2002).

For planning the future development of mountain agriculture, policy makers and conservationists need information about the effects of these changes on biodiversity. The spatial distribution of species diversity is crucial for determining the efficient commitment of agricultural subsidies and for setting conservation priorities. Furthermore, the temporal development of species diversity has to be investigated to determine the detrimental and beneficial factors which led to the present situation of biodiversity.

In this study we analyse differences between grasshopper assemblages recorded in 1981-83 and 2002-03 in the same agriculturally managed plots in a small area of the northern part of the Central Swiss Alps. Grasshoppers were chosen as indicators because, due to their specific habitat requirements, they react sensitively to changes in grassland management (Detzel 1998; Ingrisch & Köhler 1998), and are abundant enough for meaningful samples in small areas. Possible differences in the composition of the assemblages of 1981-83 and 2002-03 are related to changes in land use. It is hypothesized that the composition of grasshopper assemblages of the meadows and pastures depauperated between 1981-83 and 2002-03 due to a degradation of habitat quality caused by an intensification of the agricultural management. Because the distributions of grasshopper species as well as the agricultural management practices are related to elevation and slope in mountain regions, we assessed the effect of these two factors on the composition of grasshopper assemblages and investigated whether changes in the assemblages over time could be observed in different altitudinal zones and in habitats of different slopes.

Study area and plots

The study was conducted in the region of Grindelwald (Bern Canton) in the northern part of the central Swiss Alps. The area, reaching from the montane up to the nival zone, is characterized by a temperate-humid to cool-humid climate. A mosaic of natural and agriculturally managed vegetation characterizes the landscape.

The dominant agricultural management types are meadows and pastures. In Switzerland, cultivated mountain regions are marked by a correlation between altitude, slope and management practices. On one hand, grassland productivity and management intensity decrease with increasing altitude. This is firstly due to vegetation development, which only allows late usage of the grasslands at higher levels and secondly to the distance from the farms. On the other hand, steep slopes are grazed or mown only once a year and no fertilizer is applied. Because access to the grasslands at different altitudes depends on time, stepwise management of different altitudinal zones is practiced. In early spring, cattle graze in the lower valleys. As the vegetation period progresses, cattle are brought up step by step to higher altitudes, reaching the highest regions in July. Meanwhile, highly productive meadows at lower altitudes are mown at least twice. Less productive, unfertilised meadows at medium altitudes are mown in August and early October, while the cattle go back to the lowest regions.

According to this management system, records were repeated on twenty plots, on which Schiess (1988) recorded grasshoppers in 1981-83. They were situated at the bottom of the valley at 900 m and on the south slopes up to 2000 m a.s.l. (Tab.1). Seven are productive meadows in the lower montane zone (900-1100 m a.s.l.). Four unfertilised pre-alpine pastures (Vorsassen) and four unfertilised meadows are situated in the upper montane zone (1100-1500 m a.s.l.) and five pastures are in the subalpine zone (1500-2000 m a.s.l.). The borders of the altitudinal zones were defined according to Pfister (1984). Flat plots were found only at the bottom of the valley. Percentage slope varied from slightly sloped to steep in each zone (Tab. 1). The size of the plots was 0.2-0.3 hectares.

Grasshopper sampling

The study of Schiess (1988) in 1981-83 aimed to determine the distribution areas of grasshoppers in Grindelwald and relate them to different variables. Thus, he (1988) recorded grasshoppers only once but nearly all meadows and pastures of the valley (Tab. 1) using a transect covering the plot and its major habitat structures (shrubs, stones, open ground). The transects had a length of about 100 m and were carried out between 10 a.m. to 5 p.m. on sunny days through August to early

October when most grasshopper species can be found in the adult stage. The time spent on each plot was 10-15 minutes. Grasshopper species were identified by their individual sound and inaudible species by sight. Species were recorded on both sides of the transect up to a distance of 2 m. Grasshoppers in larval stages were ignored. Species of Tetrigidae and Gryllidae were not considered.

In 2002-03, the same method was used, but we selected 20 plots (meadows and pastures). Each plot was visited twice in 2002 during August and September and 5 times in 2003 from mid June to late August (Tab.1).

The scientific names of the grasshoppers are according to Coray & Thorens (2001).

Tab.1: Management in 2002/03, position, slope and date of recording of the 20 plots. Altitudinal zones: LM = lower montane; UM = upper montane zone; SA = subalpine zone.

Plot	Management 2002-03	Swiss grid (x/y)	Altitude m	Slope %	Date of record 1981-83	Date of record 2002		Date of record 2003				
LM1	gcc+f.	641110/165243	920	18	19.8.82	28.7	5.9	19.6	30.6	30.6	4.8	25.8
LM2	gcc+f.	644595/163538	940	0	23.8.82	23.7	22.8	20.6	30.6	14.7	7.8	21.8
LM3	gcc+f.	644525/163636	940	0	23.8.82	23.7	19.8	10.6	28.6	14.7	7.8	21.8
LM4	gcc+f.	640933/164837	975	33	19.8.83	28.7	4.9	10.6	25.6	15.7	4.8	25.8
LM5	cc+partially f.	641435/165540	1005	61	19.8.83	28.7	4.9	11.6	25.6	16.7	4.8	25.8
LM6	ccg+f.	643185/165146	1070	46	27.8.81	29.7	29.8	19.6	30.6	20.7	8.8	20.8
LM7	ccg+f.	647223/164319	1105	16	28.9.82	22.7	19.8	10.6	27.6	12.7	5.8	20.8
UM1	gg unf.	646307/164774	1210	27	1.10.81	22.7	19.8	20.6	30.6	12.7	20.7	8.8
UM2	gg unf.	644132/165227	1220	38	6.9.81	30.7	4.9	21.6	30.6	18.7	1.8	21.8
UM3	gg unf.	647237/166036	1435	35	18.8.82	24.7	20.8	21.6	2.7	14.7	7.8	19.8
UM4	c unf.	643592/165822	1440	60	8.9.83	31.7	2.9	13.6	28.6	14.7	7.8	26.8
UM5	c unf.	643411/165733	1450	54	8.9.83	31.7	2.9	12.6	28.6	19.7	7.8	26.8
UM6	c unf.	647670/165979	1470	63	5.10.81	29.7	22.8	13.6	27.6	15.7	5.8	25.8
UM7	c unf.	647139/166160	1475	30	18.8.82	24.7	22.8	12.6	27.6	14.7	7.8	19.8
UM8	gg unf.	644322/166142	1525	62	7.9.81	31.7	2.9	21.6	2.7	18.7	6.8	26.8
SA1	gg unf.	649883/166239	1620	28	19.9.82	23.7	30.8	10.6	29.6	12.7	6.8	26.8
SA2	gg unf.	648268/167078	1695	67	11.8.82	30.7	22.8	13.6	29.6	15.7	5.8	21.8
SA3	gg unf.	647797/167554	1925	48	15.9.82	30.7	30.8	12.6	25.6	15.7	5.8	20.8
SA4	gg unf.	647797/167554	1940	60	3.9.82			10.6	29.6	13.7	6.8	26.8
SA5	gg unf.	647914/167924	2000	31	15.9.82	30.7	30.8	12.6	25.6	13.7	5.8	20.8

c: cut; cc: cut twice a year; g : grazed; gg: grazed in spring and autumn; f: fertilised; unf: unfertilised

Data used in the analyses

2002-03, the sampling effort was seven times higher than in 1981-83. Ingrisch & Köhler (1998) recommend three to four records to collect 90% of grasshopper species in a given area or at least two in August and September if Tetrigidae and Gryllidae are not considered. Since only one record was realised in 1981-83, this data collection only partly fulfils this recommendation. Due to the method used by Schiess (1988), the chance of missing species was high in 1981-83. In particular, early occurring species could not be detected in 1981-83 because no records were taken before

August (Tab. 1). Furthermore, the probability of finding species in particular plots is, of course, reduced if only one record is made, particularly for species with a cryptic behaviour.

In a preliminary approach, we tried to eliminate the difference in sampling effort by a selection of single records made in 2002-03 on a comparable date or, for most equal calculated day degree sum to compare them to the single records of 1981-83. This comparison gave no meaningful results because some species found in 1981-83 in the particular plots were detected at another date in 2002-03.

To avoid confusion between ecological meaningful changes and those probably caused by the method used, we divided the analyses into two distinct parts, a so called 'absence analysis' and 'presence analysis'. In both parts of the analyses the assemblages of 1981-83 and 2002-03 were compared on a plot by plot and on a presence-absence basis.

1) Absence analysis (A)

In this analysis, we compared the species found in each particular plot in 1981-83 with the members of these species that could still be detected in the same plot in 2002-03. Therefore, we decided firstly not to consider the species exclusively found in 2002-03 (3 species) and secondly to eliminate the occurrences of species in particular plots in which they were observed only in 2002-03. Consequently, only the 18 species found in 1981-83 were considered in this analysis. Therefore, the assemblages of 2002-03, which were compared to those of 1981-83 in each plot, could only consist of the species found in 1981-83 at maximum or less. The analysis is named 'absence analysis' because some of the species found in 1981-83 were no longer found (= absent) in the assemblages of particular plots in 2002-03. One can be sure that these species no longer occurred in 2002-03 because in 2002-03 the sampling effort was seven times higher than in 1981-83. Thus, the absence of these species must have some ecological causes.

2) Presence analysis (P)

The presence analysis compares the dataset of 1981-83 with the complete set of 2002-03. Since all species found in 1981-83 were also found in 2002-03, this analysis is called 'presence analysis'. We used this analysis to describe the actual state of grasshopper distribution in 2002-03 and to supplement and rate the findings of the absence analysis.

Statistical analysis

Redundancy Analysis (RDA) and partial RDA were used to test for the influence of explanatory variables with the CANOCO 4.5 program (Ter Braak & Wiertz 1994; Ter Braak & Šmilauer 2002). The following nominal explanatory variables were introduced (with classes between brackets): time (1981-83, 2002-03), altitudinal zone (LM=lower montane zone, UM=upper montane zone, SA=subalpine zone, roughly indicating altitude and hence decreasing productivity), plot (7 in LM, 8 in UM, 5 in SA) and the cross-classification altitudinal zone by time (LM1981-83, LM2002-03,..., SA2002-03). These factors were coded as dummy variables. The 31 variables were not independent and took 22 degrees of freedom. In addition, altitude and slope of the plots were introduced as quantitative variables (Tab. 1).

Explanatory variables were made covariables to adjust for their effect on the species data. In the resulting partial ordination, their effect is eliminated. In the analysis that focuses on the time change, spatial differences were accounted for by specifying the plot dummy variables as covariables.

The total sum of squares in the species data was decomposed as in an analysis of variance table (Ter Braak & Wiertz 1994; Borcard et al. 1992).

In the ordination diagrams, the classes of nominal explanatory variables were indicated by centroids, which are average scores of the samples belonging to that particular class of explanatory variables.

Monte Carlo permutations were used to test for statistical significance. For each test, 499 permutations were performed.

Results

Species number

In 1981-83, a total of 18 grasshopper species were found in the twenty plots (Tab. 2). In the lower montane zone 12 species occurred with an average of 5.5 per plot. In the upper montane zone 16 species were observed with an average of 8.8 per plot. In the subalpine zone 11 species were found with an average of 7 per plot.

In 2002-03, a total of 21 species were found in the twenty plots (Tab. 2). All 18 species found in 1981-83 were again found in 2002-03. In the lower montane zone 16 species were present with an average of 9 species per plot. In the upper montane zone 20 species were found with an average of 14 species per plot. In the subalpine zone 15 species were recorded. The average species number here was 9.4 per plot.

Tab. 2: Grasshopper assemblages found in the 20 plots in 1981-1983 (Schiess, 1988) and in 2002-2003 near Grindelwald.

Plot ID	Lower montane zone (LM)							Upper montane zone (UM)								Subalpine zone (SA)					Occupied plots in 1981-83	Change of occurrence between 1981-2003	
	LM1	LM2	LM3	LM4	LM5	LM6	LM7	UM1	UM2	UM3	UM4	UM5	UM6	UM7	UM8	SA1	SA2	SA3	SA4	SA5		-	+
Chorthippus parallelus [A,P]	=	=	=	=	=	=	=	=	=	=	=	=	=	=	=						20	0	0
Decticus verrucivorus [A,P]	=	+	+	+	=	+	=	=	=	=	=	=	=	=	=					+	15	0	+5
Omocestus viridulus [A,P]	=	=	=	=	+	+	+	+	=	=	=	=	=	=	=					=	15	0	+4
Chorthippus biguttulus [A,P]	+	=	+	=	=	+	=	+	=	+	=	=	+	=	=						13	0	+4
Metrioptera roeselii [A,P]	=	=	=	+	=	+	+	=	=	=	=	=	=	=	=						12	0	+2
Stenobothrus lineatus [A,P]	+	=	=	=	+	+	+	+	+	+	+	+	+	+	=	=	+	=	=	+	11	0	+3
Euthystira brachyptera [A,P]	=	=	=	=	+	=	=	+	=	+	+	=	=	=	=						9	0	+4
Metrioptera brachyptera [A,P]	=	=	=	=	=	=	=	=	=	+	+	=	=	=	=					=	8	0	+4
Gomphocerippus rufus [A,P]	=	=	=	=	=	=	=	=	=	=	=	=	=	=	=	=	-				8	-1	+2
Tettigonia cantans [A,P]	=	-	=	+	=	=	+		=	=	+	=	=	=	+						7	-3	+6
Pholidoptera griseoaptera [A,P]	=				=			+	+	=		=	=	=	=						6	-1	+1
Arcyptera fusca [A,P]		-						+			-		+	=			+	=		+	5	-2	+5
Gomphocerus sibiricus [A,P]								+		-						-	=	=	=		4	-2	+1
Chorthippus dorsatus [A,P]			+				=	-		+	+	+	+				+	+			2	-1	+6
Miramella alpina [A,P]								-		+	+	-	+			-	+	+		+	2	-2	+5
Chorthippus montanus [A,P]	+							=			=		=				+				2	0	+1
Chorthippus brunneus [A,P]	+					=		+				+		+	+						1	0	+4
Stetophyma grossum [A,P]	+																+	=			1	0	+3
Tettigonia viridissima [P]	+				+	+	+	+	+		+	+	+	+	+						0	0	+8
Platycleis albopunctata [P]														+	+						0	0	+4
Omocestus rufipes [P]								+	+						+						0	0	+3
Number of species in 1981-83	6	4	4	5	8	6	6	7	10	6	9	12	9	9	9	8	8	8	7	4			
% species found again 2002-03	100	75	75	80	87.5	100	100	72	100	100	100	83	89	100	100	75	87.5	100	100	100			+75
Newly found in 2002-03	4	1	1	4	4	6	4	7	4	4	2	2	6	5	5	0	6	2	0	4		-12	

A : species of the absence analysis
P : species of the presence analysis
= : species recorded in 1981-83 and in 2002-03
- : species not found in 2002-03 as compared to 1981-83
+ : species newly recorded in 2002-03 as compared to 1981-83

Clearly, the total number of species found in 2002-2003 was higher than that of 1981-83 very probably due to the higher sampling effort. The number of species and the average number of species per plot recorded in each altitudinal zone were also higher in 2002-03 than in 1981-83. Interestingly, the proportions of species numbers and the proportions of averages between the three altitudinal zones were similar in both studies. The upper montane zone harboured most species and the plots here were the most species rich.

Analysis of the species loss between 1981-83 and 2002-03 ('absence analysis')

This analysis describes the species loss in particular plots between 1981-83 and 2002-03 because the possible sampling effort effect is eliminated by the removal of the new occurrences (+ symbols in Tab. 2).

Spatial variation:
The major variation of the 40 grasshopper assemblages (20 of 1981-83 and 20 of 2002-03) across space and time is shown in a biplot based on Redundancy Analysis (RDA) with model 'plot + year + altitudinal zone x time' (Fig. 1). For synthesizing the information, centroids of zone-by-time combinations are shown instead of the individual assemblages. The biplot shows that the composition of grasshopper assemblages in the three altitudinal zones differ distinctively from each other (Fig. 1). The arrows of altitude and slope indicate that this difference is related to these two environmental factors. The correlation coefficients between the first axis in Fig. 1 and both altitude and slope are > 0.7.

Tab. 3. Decomposition of the total variance in the species data obtained by redundancy analysis with focus on species loss ('absence analysis').

Type of analysis				Absence
Source	df	ss	ms	variance explained
Space (plot)	19	0.95	0.05	95%
Altitude	1	0.16	0.16*	17% of space
Slope	1	0.13	0.13*	13% of space
Residual	17	0.66	0.039	
Time (year)	1	0.005	0.005*	0.5%
Space x time	19	0.045	0.002	4.5%
Zone x year	2	0.004	0.002^{ns}	9% of space x time
Slope x year	1	0.006	0.006^{ns}	13% of space x time
Residual	16	0.035	0.002	
Total	39	1		100%

df = degrees of freedom; ss = sum of squares; ms = mean square;
*: $p < 0.05$ (Monte Carlo permutation); ns = not significant

Additionally, Fig. 1 shows species explaining more than 40% of the variance and therefore their occurrence in the absence dataset is associated with the displayed environmental factors.

Variance decomposition based on RDA models shows that space explains most of the total variance, i.e. 95% (Tab. 3). Altitude and slope together account for about 29% of the spatial variation. Both variables have a significant effect on grasshopper assemblages (p < 0.05) as assessed by a Monte Carlo permutation test. The importance of the two factors is stressed by their relatively high mean square values (Tab. 3).

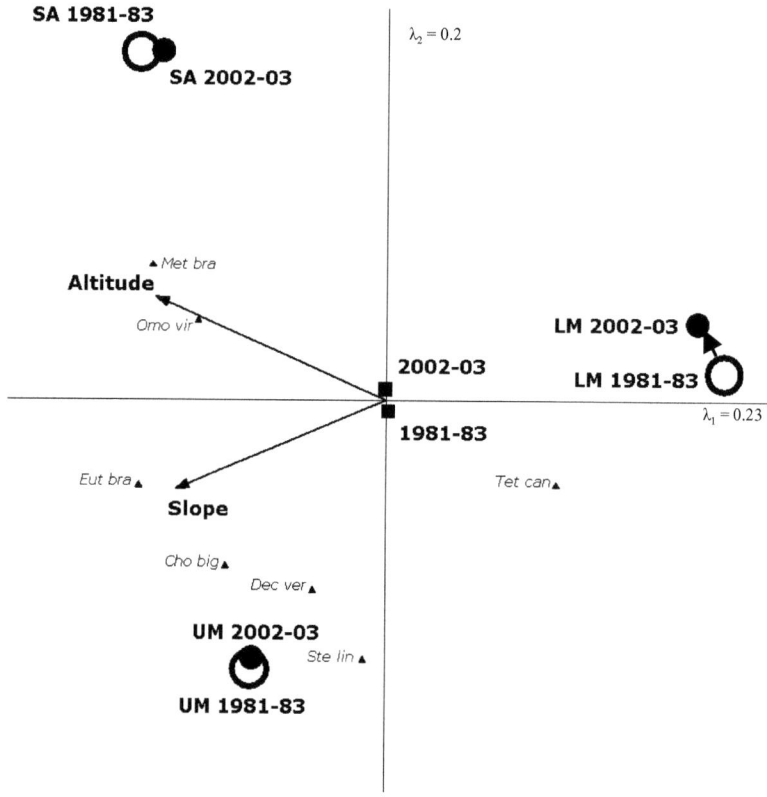

Figure 1. Ordination diagram of the spatial and the temporal variation with focus on species absence in the 40 records based on a Redundancy Analysis with model 'plot + year + zone x time'. Shown are centroids of year (1981-1983 and 2002-200 ■) and zone x year combinations (●○), as well as species which fit well (more than 40 % explained variance). Passively added to the biplot are the arrows for altitude and slope. The diagram accounts for 42 % and 45 % of the variance of the species data and the fitted species data, respectively. The first four eigenvalues are 0.23, 0.2, 0.14 and 0.1. For full species names and abbreviations LM, UM, SA see Table 2.

Temporal variation:

The temporal variation of the grasshopper assemblages between 1981-83 and 2002-03 in the absence analysis is very small (RDA ordination, Fig. 1). The amount of temporal change is largest in the lower montane zone, as indicated by an arrow connecting the centroids of 1981-83 and 2002-03 in Fig. 1. In the upper montane and in the subalpine zones the average scores of the grasshopper assemblages of 1981-83 and 2002-03 are nearly identical. Time and its interaction with space account only for 0.5 and 4.5%, respectively (Tab. 3). The interaction of zone-time (zone x year) and slope-time (slope x year) together account for 22% of the full space-time interaction.

In nine plots, one to two species disappeared (Tab. 2). In eleven plots, the complete assemblages found in 1981-83 were refound in 2002-03. The difference between the species assemblages of 1981-83 and 2002-03 is statistically significant (model: variable 'time' and covariable 'plot'; $p < 0.05$), even though it explained only 0.5% of the variance. This shows that a few, distinct species, e.g. *Acryptera fusca* and *Tettigonia cantans*, no longer occurred in some of the assemblages recorded in 2002-03 (Tab. 2). This is probably caused by slight management changes. However, this change over time showed no spatial pattern. It was independent of the altitudinal zones (model: variable 'time x altitudinal zone' and covariable 'plot' and 'time') as indicated by the Monte Carlo permutation test (Tab. 3). This means that the absences are not exclusively occurring in one particular altitudinal zone. The same can be concluded for assemblages in plots of different steepness (model: variable 'slope x time' and covariable 'plot' and 'time'). Absences of species are observed in flat and slightly sloped as well as in steep plots.

The RDA ordination of Fig. 2 shows the loss of species between the assemblages of 1981-83 and 2002-03 and its independence of the three altitudinal zones. Species in the left side of the ordination are those locally no longer found in 2002-03. Species in the top of the diagram, e.g. *Miramella alpina*, locally no longer occurred in certain plots at high altitudes, species at the bottom of the ordination, e.g. *Tettigonia cantans* no longer occurred at low altitudes. The first two axes display 92% of the observed variance. The first axis represents the difference of the grasshopper assemblages between 1981-83 and 2002-03 indicated by the position of the time centroids 1981-83 and 2002-03. The eigenvalue of the first axis ($\lambda = 0.006$) indicates that the differences between the assemblages of 1981-83 and 2002-03 are very small. The position of the zone-time centroids in the ordination shows that the variable 'time' explains similar variances in each altitudinal zone.

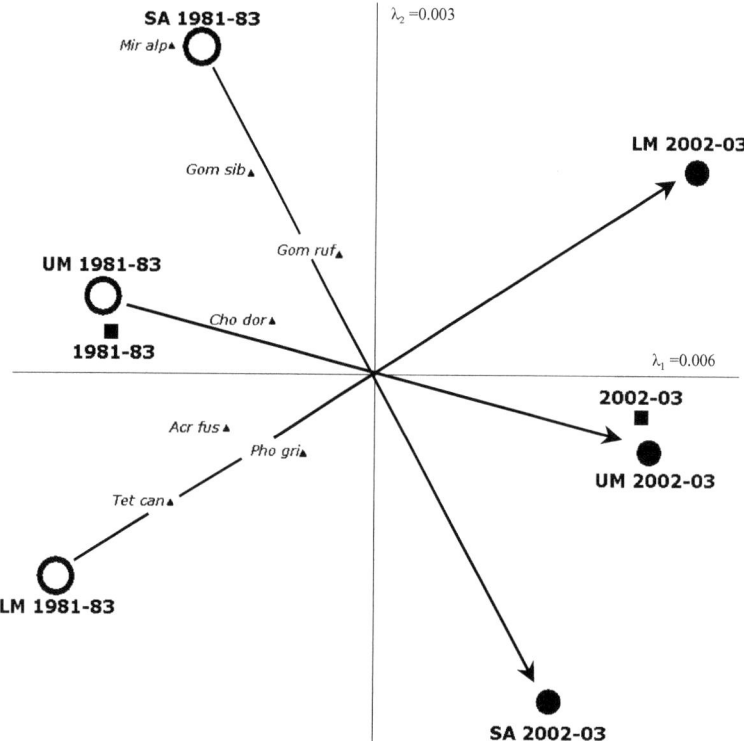

Figure 2. Ordination diagram of the species loss over time in more detail than in Fig. 1. Shown are centroids of zone x year combinations (●○) and species that locally disappeared. The species loss between 1981-1983 (■) and 2002-2003 (■) is emphasized by connecting the year scores for each zone. The diagram accounts for 16 % and 85 % of the residual variance of the species data and the fitted species data, respectively. The first four eigenvalues are 0.006, 0.003, 0.001 and 0.014. For full species names and abbreviations LM, UM, SA see Table 2.

Analysis of additionally found species in 2002-03 in the particular plots ('presence analysis')

In this analysis, the differences between the assemblages of 1981-83 and 2002-03 are mainly due to the different sampling efforts in the two study periods. In almost all plots, more species were found in 2002-03 than in 1981-83 because the chance of finding a species was seven times higher (Tab. 2). Thus, an interpretation of these temporal differences in terms of habitat changes is difficult. However, the more detailed grasshopper recordings of 2002-03 in the presence analysis allow a better spatial interpretation of the present species distribution than the absence analysis.

Spatial variation:

The spatial variation of the grasshopper assemblages among the altitudinal zones is still prominent and related to altitude and slope (Fig. 3). In this analysis, space explained also most of the total variation, i.e. 67% (Tab. 4). Altitude and slope still have a significant effect on the composition of grasshopper assemblages and together account for about 37% of the spatial variance. The mean square values are relatively high and again stress the importance of the two factors (Tab. 4).

Obviously, the basic differences between the assemblages in the three altitudinal zones were already found with the single transect carried out in 1981-83 (Fig. 3). The distances between the centroids of the altitudinal zones of 1981-83 have nearly the same length as those between the centroids of 2002-03. But due to the seven times higher sampling effort in 2002-03, the distribution of the grasshoppers are described in more detail than in 1981-83.

In Fig. 3 species explaining more than 40% variance are shown. *Pholidoptera griseoaptera* and *Tettigonia cantans* are characteristic for assemblages of the montane zones with a preference for the lower zones of these. *Chorthippus biguttulus* and *Euthystira brachyptera* occur in the steepest plots. *Metrioptera brachyptera* occurs at higher altitudes, mainly at the border between the upper and the subalpine zones.

Tab. 4: Decomposition of the total variance in the species data obtained by redundancy analysis with focus on the species additionally found in particular plots ('presence analysis).

Type of analysis			Presence	
Source	df	ss	ms	variance explained
Space (plot)	**19**	**0.67**	**0.035**	**67%**
Altitude	1	0.14	0.14*	21% of space
Slope	1	0.11	0.11*	16% of space
Residual	17	0.42	0.025	
Time (year)	**1**	**0.054**	**0.054***	**5.4%**
Space x time	**19**	**0.276**	**0.006**	**27.6%**
Zone x year	2	0.043	0.021*	16% of space x time
Slope x year	1	0.015	0.015^{ns}	5% of space x time
Residual	16	0.19	0.012	
Total	**39**	**1**		**100%**

df = degrees of freedom; ss = sum of squares; ms = mean square;
*: $p < 0.05$ (Monte Carlo permutation); ns = not significant

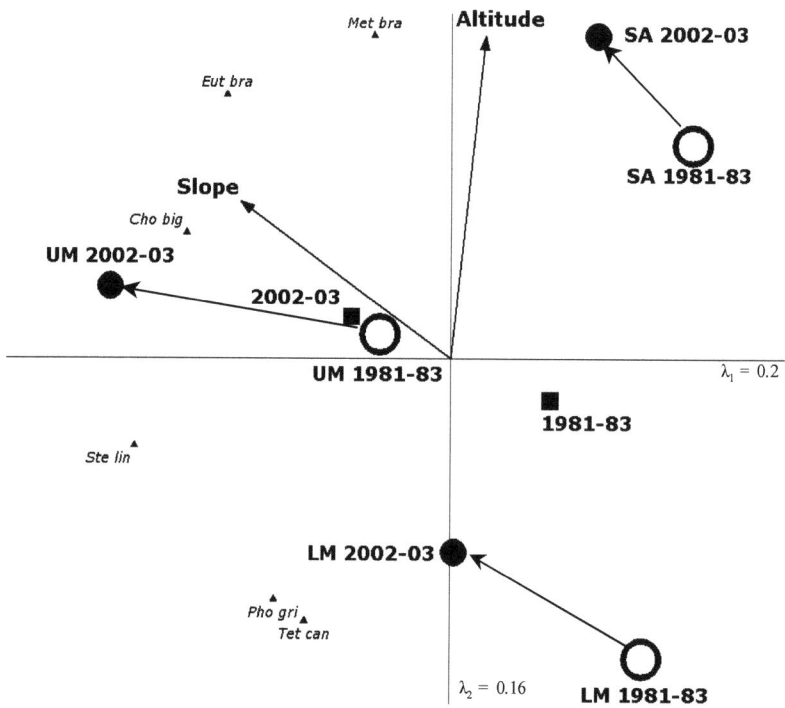

Figure 3. Ordination diagram of the spatial and the temporal variation with focus on newly occurring species in the 40 records based on a RDA with model 'plot + year + zone x time'. Shown are centroids of year (1981-1983 ■ and 2002-2003 ■) and zone x year combinations (● ○), as well as species which fit well (more than 40 % explained variance). Passively added to the biplot are the arrows for altitude and slope. The diagram accounts for 41 % and 48 % of the variance of the species data and the fitted species data, respectively. The first four eigenvalues are 0.21, 0.16, 0.12 and 0.06. For full species names and abbreviations LM, UM, SA see Table 2.

Temporal variation:

The temporal variation of grasshopper assemblages in the different altitudinal zones is indicated by the length of the arrows connecting the 'zone-by-time' centroids of 1981-83 and 2002-03 (Fig. 3). Thus, large differences exist between the datasets of 1981-83 and 2002-03. The reason for these differences is that more species and individuals were recorded in 2002-03 compared to 1981-83 due to the larger sampling effort.

Time and its interaction with space explain 5.4% and 27.6% of the total variance, respectively (Tab. 4). Slope-time (slope x year) and zone-time (zone x year) interaction together account for about 21% of the full space-time interaction.

To test the change over time and its dependence on altitudinal zone and slope, the same model as in the absence analysis was used. As expected, the Monte Carlo permutation test again indicated a significant difference in the grasshopper assemblages of 1981-83 and 2002-03 ($p < 0.05$). In 18 plots between one to seven additional grasshoppers were recorded (Tab. 2). In contrast to the absence analysis, this difference was dependent on the altitudinal zone ($p < 0.05$). The largest increase of newly found species in 2002-03 was observed in the plots of the upper montane zone. As in the absence analysis, the difference of the assemblages in the two studies was independent of slope.

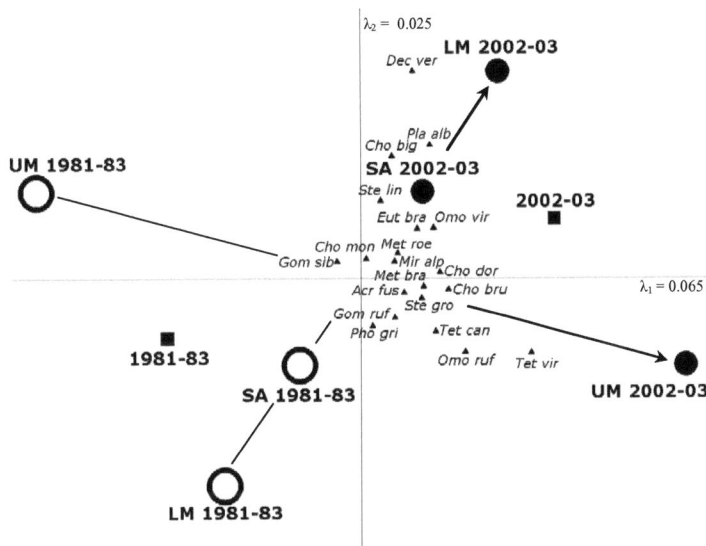

Figure 4. Ordination diagram of the increase of frequency of the species over time in more detail than in Fig. 3. Shown are centroids of zone x year combinations (●○) with focus on species newly found in 18 of the investigated plots. The increase of species frequency between 1981-1983 and 2002-2003 (■) is emphasized by connecting the year scores for each zone. The diagram accounts for 27 % and 92 % of the residual variance of the species data and the fitted species data, respectively. The first four eigenvalues are 0.06, 0.02, 0.007 and 0.04. For full species names and abbreviations LM, UM, SA see Table 2.

The difference in grasshopper assemblages of 1981-83 and 2002-03 and its dependence on the altitudinal zone is shown with the RDA ordination of Fig. 4. The first two axes display 93% of the variance accounting for the model 'time + altitudinal zone' with covariable 'plot' and 'time'. As in the absence analysis, the first axis represents the difference between the 1981-83 and 2002-03 species compositions. The eigenvalue of the first axis is higher in this model ($\lambda = 0.065$) than in the absence analysis, indicating that the variable 'time' now better differentiates the assemblages than the absence analysis. The position of the centroids shows that the difference between the assemblages of 1981-83 and 2002-03 was larger in the upper montane zone than between those of the lower and subalpine zones. The 19 species placed on the right side of the second axis occupy more plots in 2002-03 than in 1981-83, while on the left side, *Gomphocerus sibiricus* is the only species occupying more plots in 1981-83.

The arrows between the centroids of the lower montane and the subalpine zones nearly point in the same direction (Fig. 4). This indicates that some of the species newly found in the lower montane zone were also newly found in the subalpine zone, e.g. *Decticus verrucivorus* and *Chorthippus biguttulus* (Tab. 2).

Discussion

Spatial variation

The results show that the composition of grasshopper assemblages changes along the altitudinal gradient.

This finding is consistent with the results of other studies on grasshoppers and further arthropods (Wettstein & Schmid 1999; Burla & Bächli 1991). At low altitudes, species that are adapted to warm temperatures are common, e.g. *Platycleis albopunctata* and *Pholidoptera griseoaptera*. At higher altitudes they are replaced by typical mountain species such as *Gomphocerus sibiricus* and *Metrioptera brachyptera*. This pattern corresponds to that described in Detzel (1998) and Nadig (1986).

The upper montane zone is the most species rich zone because here lowland species occur at their upper border and mountain species at their lower border (Tab. 2). The predominant land use types, i.e. extensively managed pastures and hay meadows, are another reason for the high species number and the high average species number per plot in this zone. The management pressure in these meadows and pastures is low and many grasshopper species profit from the fact that these grasslands remaining undisturbed over a long time during the season (see Detzel 1998; Gerstmeier & Lang 1996; Dolek 1994; Bosshard *et al.* 1988).

As stressed by Ingrisch & Köhler (1998) and Schmidt (1983), fewer grasshopper species can be found in fertilised and frequently mown meadows. In our study, the lowest average number of species per plot was found in the lower montane zone, where the most intensively managed meadows occur.

It is surprising that the effect of slope on grasshopper assemblages is almost as strong as the well-known effect of altitude. A probable reason is that steep south slopes are often dryer than flat plots because the evapotranspiration and runoff of water increases with the steepness of the slopes. The occurrence of many grasshoppers is related to the dryness of the habitat (Detzel 1998). For example, *Chorthippus brunneus* occurs only in dry sites, whereas *Chorthippus montanus* and

Stethophyma grossum only occur in moist grasslands (Detzel 1998). Another reason for the effect of slope is that the management pressure decreases with increasing slope because steep plots are difficult to manage. Thus, steep grasslands are attractive for species sensitive to high management pressure, such as *Chorthippus biguttulus* and *Stenobothrus lineatus* (Detzel 1998).

Temporal variation

Abandonment of grassland management plays a minor role in Grindelwald (M. Peter, oral communication). We assume that socio-economic conditions (e.g. tourism), subsidies for agriculture and a high esteem for local agricultural traditions enable the farmers to maintain their traditional management forms.
In the absence analysis, a very slight (0.5% variance) but significant loss ($p < 0.05$) of grasshoppers as compared to the assemblages of 1981-83 was observed. An interesting fact is that this loss did not occur exclusively in a particular altitudinal zone or in plots of a particular slope. Hence, the management intensity has not increased in any altitudinal zone. Such an intensification could have been expected on flatter slopes due to technical improvements of mowing and other farm machinery.
The detected, small loss between 1981-83 and 2002-03 cannot be attributed to a slight intensification over all plots. In the plots where species disappeared, the assemblages of 2002-03 still have species from 1981-83 sensitive to intensive management, which would have also been harmed by intensification, e.g. *Stenobothrus lineatus* and *Euthystira brachyptera* (Detzel 1998). Additionally, the presence analysis shows that almost all species that have locally disappeared since 1981-83 occupy many other similar plots in 2002-03.
All this means that it is more likely that in the different plots different habitat changes, such as a removal of a large shrub or increased shelter through newly grown trees, caused the extinction of a particular species between 1981-83 and 2002-03 rather than a general change of land use management.
In 2002-03, many additional grasshoppers were found in the investigated plots. Of course, most of these species were newly found due the higher sampling effort: the chance to find rare species, such as *Platycleis albopunctata*, was seven times higher in 2002-03 than in 1981-83. *Tettigonia cantans* and *Tettigonia viridissima* were found in many additional plots in 2002-03. Both species have a cryptic behaviour during the day and they start stridulating in the evening. A high sampling effort raises the chance of finding these two species. Furthermore, species occurring early in the season such as *Omocestus rufipes* were additionally found because records were also made before August in 2003.

Another reason for the high amount of additionally found species could be the optimal climatic conditions in the hot summer of 2003, which probably enabled some species to occupy habitats at higher altitudes, e.g. *Stenobothrus lineatus* and *Chorthippus montanus*.

Combining the results of the absence and the presence analyses allows us to put the observations in broader perspective. Meadows and pastures are important habitats for grasshoppers in the Grindelwald region. Due to the economic situation of the farmers in mountain regions, an increase of the grassland productivity with problematic consequences for biodiversity might have been necessary to maintain their living (Bernegger 1990). But the high grasshopper diversity found in these grasslands in 2002-03 indicates habitats of high quality and a very appropriate management from the grasshopper's perspective. The presence of *Euthystira brachyptera* in nearly all plots on the south slopes in 2002-03 rather suggests very low management intensity because this species typically occurs in unmanaged grasslands (Detzel 1998).

It must be assumed that the loss of species is underestimated in our study. It is probable that some species disappeared in particular plots since 1981-83, but they were missed in 1981-83 due to the low sampling effort. About these species, we have unfortunately no information. However, the presence analysis determines the species that potentially were missed in 1981-83 in the particular assemblages. If all these species were just not recorded in the particular meadows and pastures in the past, it can be concluded that the assemblages were similarly rich in grasshopper species in 1981-83 as they are in 2002-03.

The small species loss between 1981-83 and 2002-03 and the high amount of species refound in the investigated plots suggest that the habitats and their management have hardly changed over the last 20 years. These findings show that the agriculture in Grindelwald can preserve grasshopper diversity as long as it is practiced as in the last two decades.

Conclusions

The Grindelwald region is a hotspot for grasshopper diversity. With 21 species the twenty plots harbour about 20% of the Swiss Orthoptera fauna, Gryllidae and Tetrigidae not included (Nadig & Thorens 1994). Of these 21 species found in 2002-03 five are rated endangered and two are strongly endangered according to the Red List of northern Switzerland (Thorens & Nadig 1997).

Our study showed that up to now the agricultural management could ensure the living conditions of all grasshopper species already found in 1981-83. This stresses how crucial it is to include the agricultural practices into nature conservation plans. Special focus should be given to the extensively managed pastures and meadows in the upper montane zone.

In most parts of the Alps, abandonment and intensification of agriculture are two major factors endangering grassland biodiversity. So far, both factors do not threaten grasshopper populations in Grindelwald.

Therefore, the Grindelwald region is an important example on how the mountain agriculture can be maintained without increasing grassland productivity and how the agricultural management can prevent a decline of grasshopper diversity.

Acknowledgements

We are grateful to the farmers in the region of Grindelwald for allowing us to work on their land and to Heinrich Schiess for using his data. Further, we thank Catherine Palmer for the language corrections. This study was financed by Swiss National Science Foundation (grant 4048-064405).

References

Bernegger, U., Cavegn, G., Meyer, L. and Rieder, P. 1990. Existenzfähige Bergbauern- betriebe als Voraussetzung für die Nutzung von Grenzertragsböden und einer gesicherten Besiedlung in nicht-touristischen Bergdörfern. Nationales Forschungsprogramm 22, Bericht 34, Liebefeld-Bern.

BFS/Geostat 1997. Swiss land-cover statistics 1992/97. Swiss Federal Office for Statistics, Bern.

Borcard, D., Legendre, P. and Drapeau, P. 1992. Partialling out the spatial component of ecological variation. Ecology 73: 1045-1055.

Bosshard, A., Andres, F., Stromeyer, S. and Wohlgemuth, T. 1988. Wirkung einer kurzfristigen Brache auf das Ökosystem eines anthropogenen Kleinseggenrieds – Folgerungen für den Naturschutz. Berichte des Geobotanischen Instituts ETH, Stiftung Rübel, Zürich, 54: 181-220.

Burla, H. and Bächli, G. 1991. A search for pattern in faunistical records of drosophilid species in Switzerland. Zeitschrift für Zoologische Systematik und Evolutionsforschung, 29, 176-200.

Coray, A. and Thorens, P. 2001. Heuschrecken der Schweiz; Bestimmungsschlüssel. Fauna Helvetica, CSCF and SEG, Neuchâtel, 5: 1-236.

Detzel, P. 1998. Die Heuschrecken Baden-Württembergs. Ulmer, Stuttgart.

Dietl, W. 1996. Zur pfleglichen Nutzung der Wiesen und Weiden im Berggebiet. Montagna 6: 17-18.

Dolek, M. 1994. Der Einfluss der Schafbeweidung von Kalkmagerrasen in der Südlichen Frankenalb auf die Insektenfauna (Tagfalter, Heuschrecken). Agrarökologie 10, Haupt, Bern.

Erhardt, A. 1985. Diurnal Lepidoptera: sensitive indicators of cultivated and abandoned grassland. Journal of Applied Ecology 22: 849-861.

Garcia-Ruiz, J., Lasante, T., Ruiz-Flano, P., Ortigosa, L., White, S., Gonzales, C. and Marti, C. 1996. Land-use changes and sustainable development in mountain areas: a case study in the Spanish Pyrenees. Landscape Ecology 11: 267-277.

Gerstmeier, R. and Lang, C. 1996. Beitrag zu Auswirkung der Mahd auf Arthropoden. Zeitschrift für Ökologie und Naturschutz 5: 1-14.

Gotsch, N., Flury, C., Kreuzer, M., Rieder, P., Heinimann, H.R., Mayer, A.C. and Wettstein, H.-R. 2004. Land- und Forstwirtschaft im Alpenraum - Zukunft im Wandel. Schriftenreihe Nachhaltige Land- und Forstwirtschaft im Alpenraum 8, Wissenschaftsverlag Vauk, Kiel.

Hintermann, U., Weber, D., Zangger, A. and Schmill, J. 2002. Biodiversity Monitoring in Switzerland, BDM-Interim Report, Swiss Agency for Environment, Forests and Landscapes SAEFL, 342.

Ingrisch, S. and Köhler, G. 1998. Die Heuschrecken Mitteleuropas. Westarp Wissenschaften, Magdeburg.

Körner, C. 1995. Alpine plant diversity: a global survey and functional interpretations. In: Chapin, F.S. III, Körner, C. (Eds), Arctic and alpine biodiversity: patterns, causes and ecosystem consequences. Springer, Berlin: pp. 45-62.

Nadig, A. 1986. Ökologische Untersuchung im Unterengadin. Heuschrecken (Orthoptera). Ergebnisse der wissenschaftlichen Untersuchung im Schweizerischen Nationalpark, Chur, 12: 103-176.

Nadig, A. and Thorens, P., 1994. Rote Liste der gefährdeten Heuschrecken der Schweiz. In: Duelli, P. (Eds), Rote Listen der gefährdeten Tierarten der Schweiz, Bundesamt für Umwelt, Wald und Landschaft, Bern: pp. 66-68.

Pfister, H. 1984. Grünlandgesellschaften, Pflanzenstandort und futterbauliche Nutzungsvarianten im montan-subalpinen Bereich. Schlussbericht des Schweizerischen MAB-Programm Nr. 7, Bundesamt für Umweltschutz, Bern.

Schiess, H. 1988. Wildtiere in der Kulturlandschaft. Schlussbericht des Schweizerischen MAB-Programm Nr. 35, Bundesamt für Umweltschutz, Bern.

Schmidt, G. H. 1983. Acrididen (Insecta: Saltatoria) als Stickstoffzeiger. Veröffentlichungen der Deutschen Zoologischen Gesellschaft: 65-68.

Ter Braak, C.J.F. and Šmilauer, P. 2002. CANOCO Reference Manual and CanoDraw for Windows - User's Guide: Software for Canonical Community Ordination, Version 4.5. Biometris, Wageningen and České Budějovice.

Ter Braak, C.J.F. and Wiertz, J. 1994. On the statistical analysis of vegetation change: a wetland affected by water extraction and soil acidification. Journal of Vegetation Science 5: 361-372.

Theurillat, J.-P., Aeschimann, D., Küpfer, P. and Spichiger R. 1995. The higher vegetation units of the Alps: In: Géhu, J.M. (Eds.), Colloques Pythosociologique 23: 189-239.

Thorens, P. and Nadig, A. 1997. Atlas de distribution des Orthoptères de Suisse. Documenta Faunistica Helvetiae, CSCF, Neuchâtel, 16: 1-236.

Wettstein, W. and Schmid, B. 1999. Conservation of arthropod diversity in montane wetlands: effect of altitude, habitat quality and habitat fragmentation on butterflies and grasshoppers. Journal of Applied Ecology 36: 363-373.

Zoller, H., Bischof, N., Erhardt, A. and Kienzle, U. 1984. Biocoenosen von Grenz- ertragsflächen und Brachland in der Berggebieten der Schweiz; Hinweise zur Sukzession, zum Naturschutzwert und zur Pflege. Phytocoenologia 12: 373-394.

III. Influence of grassland management, altitude and slope on Orthoptera and diurnal Lepidoptera communities in two Swiss Alpine valleys

M. Hohl[1], A. Gigon[2], A. Erhardt[3] and P. Jeanneret[1]

[1]Agroscope FAL Reckenholz, Swiss Federal Research Station for Agroecology and Agriculture, Reckenholzstr. 191, CH-8046 Zurich, Switzerland
[2]Institute of Integrative Biology, Swiss Federal Institute of Technology, ETH Zurich, CH-8092 Zurich, Switzerland
[3]Department of Integrative Biology, University of Basel, St. Johanns-Vorstadt 10, CH-4056 Basel, Switzerland

Abstract

Traditionally cultivated grasslands in the Alps are known to be species rich in Orthoptera and Lepidoptera. Due to the ongoing abandonment or the intensification of management of extensively managed grasslands across the Alpine arc these grasslands are decreasing and have become of high conservation interest. Therefore, we studied Orthoptera and Lepidoptera in intensively and extensively managed meadows and in lightly grazed pastures on the south facing slopes of two valleys in the Swiss Alps. We analysed the influence of the different management types, the altitude and the slope on the diversity of both taxa.

In total, we observed 28 Orthoptera and 101 Lepidoptera species. These are about 30% of the Swiss Orthoptera and 46% of the butterfly fauna. 11 Orthoptera and 38 Lepidoptera are currently on the Red List of Northern Switzerland. The number of Orthoptera species did not significantly differ among the management types, while the number of Lepidoptera species was lower in the intensively managed meadows. The species composition of the investigated communities was significantly influenced by the type of grassland management, altitude and the slope. The species diversity was highest in extensively managed meadows and lightly grazed pastures. The response of Orthoptera to management was smaller than that of Lepidoptera.

The high topographic variability and the variety of the management forms in the valleys are a major source of species diversity. The Orthoptera and Lepidoptera communities in the grasslands on the south facing slopes are a mixture of lowland and mountain species as well as species with distinct microclimatic requirements which they find in grasslands of different steepness. The study showed that agriculturally managed grasslands of the Alps are of high conservation value. Extensive mowing and light grazing are the best management practices for maintaining the diversity of the two taxa and, thus a key for successful conservation of the Orthoptera and Lepidoptera fauna in Alpine valleys. However, the management practices, altitude and slope are intercorrelated. Therefore, in the Alps, it is difficult to define general conservation recommendations for meadows and pastures.

Keywords: biodiversity, meadows and pastures, habitat heterogeneity, insect conservation, mountain agriculture

Introduction

In the Alps, agriculturally managed grasslands are important habitats for Orthoptera and Lepidoptera and other insects (Ingrisch & Köhler 1998; Lepidopterologen-Arbeitsgruppe 1991; Erhardt 1995; Oertli et al. 2005). Some of these grasslands were created by human activities over the last Millennia (Ellenberg 1996). Up to recently, these grasslands were managed relatively extensively due to the difficult management conditions in the rough terrain and climate in Alpine regions (Bätzing 2003). However, the cultivated landscapes are under growing economical pressure due to the increasing costs of labour and the decreasing price of agricultural products in the liberalising markets (Bätzing 1996). In many parts of the Alps, the grassland with low yield and those difficult to manage are often abandoned due to the required rationalisation in agriculture (Internationale Alpenschutzkomission CIPRA 2001). Consequently, they become overgrown by shrubs and trees leading to a decrease of grassland diversity in the long term (e.g. Erhardt 1985a, b: Bischof 1984; Balmer & Erhardt 2000). Parallel to the abandonment, the sites that are easy to manage are likely to be more intensively managed, i.e. higher fertiliser application, more frequent mowing and increased number of livestock (Oertli et al. 2005). The intensification of grassland management usually leads to trivialised plant communities (Pfister 1984; Dietl 1996) and a decreased insect diversity (e.g. Erhardt 1985a, b; Kiser 1987; Kruess & Tscharntke 2002).

In the Swiss Alps, the cultivated grassland area remained more or less constant over the last decades (Internationale Alpenschutzkomission CIPRA 2001). At present, in most valleys a step-wise management following the altitude is practiced. Farmers start using meadows with livestock at the valley bottom in spring and follow the vegetation development through the season up to the top of the mountains. At the valley bottom after grazing, the meadows are fertilised and usually mown twice during the summer while at high altitudes some are mown only once. On steep slopes, the grasslands are usually extensively grazed or some are mown once.

This investigation focuses on Orthoptera and Lepidoptera (Rhopalocera, Hesperiidae, Zygaenidae) in two Alpine valleys; Grindelwald and Tavetsch. The reasons for choosing these valleys is that their grasshopper and diurnal Lepidoptera fauna is well known because it has been already studied by Schiess (1988), and Erhardt (1985a, b).

Further reasons for choosing these taxa are that they react sensitively to different management practices (Erhardt 1985a, b; Detzel 1998), to altitude and to slope. Grasshoppers and butterflies occur in relatively large numbers of species and individuals and are in general relatively easy to determine.

Many species of both insect groups are very mobile and often need several ecosystems types for their survival. Their presence (or absence) can thus provide information on the whole landscape.

Butterflies and to a lesser degree also Orthoptera are charismatic insect groups. They are of high conservation value as documented by the large number of species which are on the Red Lists. Results obtained for both taxa are thus of great value in discussions and decisions regarding conservation not only of the species themselves, but also concerning the ecosystems in which they occur and the whole landscape.

Methods

Study areas

The study was carried out in Grindelwald and Tavetsch, two valleys in the Swiss Alps (Fig. 1). According to Gonseth *et al.* (2001) the two case study areas belong to different bio-geographical regions. Grindelwald is in the Northern Alps where the climate is characterized as temperate-humid to cool-humid (Pfister 1984). In the Central Alps at Tavetsch, the climate is temperate-dry (Erhardt 1985b). The valley bottom of Grindelwald is at about 900 m a.s.l. and in the Tavetsch at about 1300 m.

In both areas, a step-wise management is practiced, resulting in a mosaic of meadows and pastures of different management intensity along the altitudinal gradient. Up to the timberline, meadows and pastures in both regions are mixed with shrubs and forests of different ages.

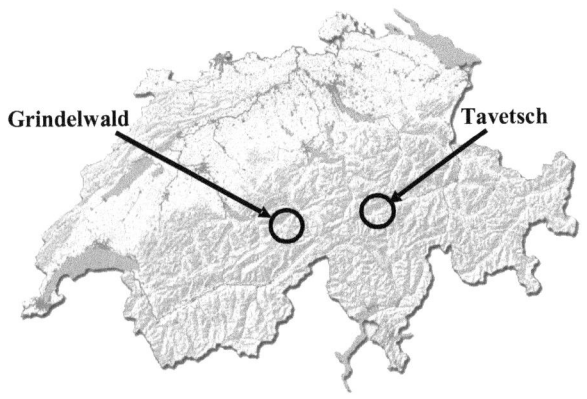

Figure 1: The study areas in the Swiss Alps. The linear distance between the two areas is about 57 km.

Study sites

In each valley 18 sites were surveyed. The sites had an area of about 2500 m² each and all were at the bottom of the valley and on south-facing slopes. In each region, three different grassland management types were studied; six sites were intensively managed meadows, six were extensively managed meadows and six were lightly grazed, i.e. they were grazed about 1-2 weeks in spring and a few days in autumn by cattle or sheep (Tab. 1). The slopes of the sites varied between 0 and 90 % and they were at different elevations (Table 1). In Grindelwald, the sites were situated between 900 and 1700 m.a.s.l. In the Tavetsch they were between 1350 and 1830 m.a.s.l.

Table 1: Management, altitude and percentage slope of the surveyed grasslands on the south slope of the Grindelwald and Tavetsch valleys.

Management type		Management	Grindelwald		Tavetsch	
			Altitude m a.s.l.	Slope %	Altitude m a.s.l.	Slope %
IM	Intensively managed meadows	2-3 x mown per year, fertilised	900-1100	0-46	1350-1550	0-36
EM	Extensively managed meadows	1 x mown per year, unfertilised	1000-1500	30-60	1500-1800	0-60
PA	Pastures	2 x grazed per year, unfertilised	950-1550	20-60	1450-1750	35-90

Orthoptera surveys

Orthoptera were recorded along two transects of 50 m covering the study site and its major habitat structures (shrubs, stones, open ground). Species were recorded on both sides of a transect up to a distance of 2 m and were identified by their individual sound or inaudible species by sight.
The time spent on each transect was 5-8 minutes and they were carried out between 10 am. and 5 pm. (Central European summer time) on sunny days.
In 2002, the transects were carried out once in early August and once in early September. In 2003, in each site five transects were carried out between mid June and mid September.
Species of Tetrigidae were not considered.
Nomenclature follows Coray & Thorens (2001).

Butterfly surveys

For recording butterflies we used the same method as Erhardt (1985a). The details of the method are given in Erhardt (1985a). In summary, an area transect was conducted, i.e. the sites were patrolled in a serpentine pattern and butterflies were identified and counted within a corridor of 5 m. Multiple counting of individuals of fast flying species can not be completely avoided using this method. However, the resulting error is assumed to be the same in all sites (Balmer & Erhardt 2000).
Transects were carried out between 10 am and 5 pm (Central European summer time) under the weather conditions defined by Erhardt (1985a), i.e. bride sunshine, a temperature minimum above 17 C° and low wind speed were considered. A single transect took about 20 minutes on average.
The butterfly surveys were carried out between mid May and the end of September. This period is optimal for butterfly recordings at these elevations (Erhardt 1985a).
In 2002, in each site four transects were carried out and in 2003 six, i.e. transects were repeated every 3 to 4 weeks.
All Rhopalocera, Hesperiidae and Zygaenidae, were recorded.
Due to difficulties in distinguishing between very similar butterfly species some species were pooled and considered as one species in the statistical analyses (*Colias alfacariensis/C. hyale, Erebia meolans/E. oeme, Erebia melampus/E. sudetica, Speyeria aglaia/Fabriciana adippe/F. niobe, Adscita alpina/A. geryon*).
Nomenclature follows Leraut (1997).

Classification of the abundances

For eliminating small, ecologically less meaningful differences among the species' abundance in the different sites, semi-quantitative scales of abundances were applied for the statistical analyses (see Erhardt 1985a).

The grasshoppers' abundances per transect were assessed using a scale of 5 classes (Table 2).

The highest value assessed per species, site and transect in 2002 or 2003 was used for the further analyses of the grasshopper communities.

Table 2: Classification of the abundances of Orthoptera species recorded during a single transect

Abundance class	Number of individuals per transect
1	1 individual per transect
2	2-10 individuals
3	11-20 individuals
4	21-40 individuals
5	> 40 individuals

The numbers of butterfly individuals counted per species, study site and year were added up and transformed into the semi-quantitative scale shown in Table 3 (see also Erhardt 1985a). For the statistical analyses, the highest class of abundance per species, site and year was used for the analysis of the butterfly communities.

Table 3: Classes of abundance according to Erhardt (1985a). Flight period = recording time of one year between Mai and September; maximum appearance = number of individuals counted in a single transect at one date.

Abundance class	Number of individuals per study site
1	1 individual per flight period
2	2-4 individuals per flight period
3	5-10 individuals per flight period
4	> 10 individuals per flight period, max. appearance <10 individuals
5	max. appearance 10-40 individuals
6	max. appearance 41-100 individuals

Statistical methods

The two study areas were situated at different elevations and they belonged to two bio-geographical regions with different climates and landscape situations. Therefore, we analysed the effects of management, slope and altitude in each study area separately.

For analysing the influence of the management on species numbers we used generalised nonlinear models. Firstly, we tested whether the interactions of altitude, slope and management have a significant effect on the species number in the investigated grasslands applying the homogeneity of slope model. Because no interaction was significant, we subsequently switched to generalised nonlinear ANCOVA models (Waldχ^2 statistic) and assumed Poisson errors and log link functions (McCaullagh & Nelder 1989). Altitude and slope were introduced as covariates, the management being the predictor variable. These analyses were performed with the program Statistica 6.1.

Redundancy Analysis (RDA) and partial RDA (Ter Braak & Prentice 1988) were used for quantifying the impact of altitude, slope and of the management [i.e. intensively managed meadows (IM), extensively managed meadows (EM) and pastures (PA)] on the species composition of grasshoppers and butterflies communities. Variation partitioning was performed, i.e. in partial RDAs, explanatory variables were made covariables to control for their effect on the species data (Borcard *et al.* 1992). In the resulting partial ordination, their effect is eliminated.

Based on Principal Component Analyses ordination diagrams of the species data were plotted. Altitude and slope were added passively in these ordinations. In the diagrams, we presented only those species that are highly influencing the ordination, i.e. the total fraction of the variance of a fitted species to the current environmental variables was at least 50% or higher (Ter Braak & Šmilauer 2002).

Monte Carlo permutations were used to test for significance. For each test, 499 permutations were performed. The multivariate analyses were carried out using the program CANOCO 4.5 (Ter Braak & Šmilauer 2002).

Indicator species for the management types

For assessing species significantly associated with a particular management type within a study area we used the indicator value method according to Dufrène & Legendre (1997). The indicator value (IndVal) is a combination of a species specificity and fidelity. Specificity of a species was at its maximum when the individuals of a species were observed in only one of the three management types. Fidelity was at its maximum when it occurred on all sites of a particular management type. The indicator value is 100% when specificity and fidelity are at their maximum. Species with high indicator value for a particular management type might probably suffer most if that particular management is not applied any more by the farmers. (Oertli et al. 2005).

The indicator values were calculated with the program INDVAL (Dufrène & Legendre 1997). To test for significance, Monte Carlo permutations were used. For each test, 499 permutations were carried out.

Results

Number of Orthoptera and Lepidoptera species

In total, 28 Orthoptera were found in the 36 sites (Fig. 2), 21 in Grindelwald and 22 in Tavetsch. Seven species were exclusively found in Grindelwald and six exclusively in Tavetsch. In general, the number of species per management type was higher in Grindelwald than in Tavetsch.

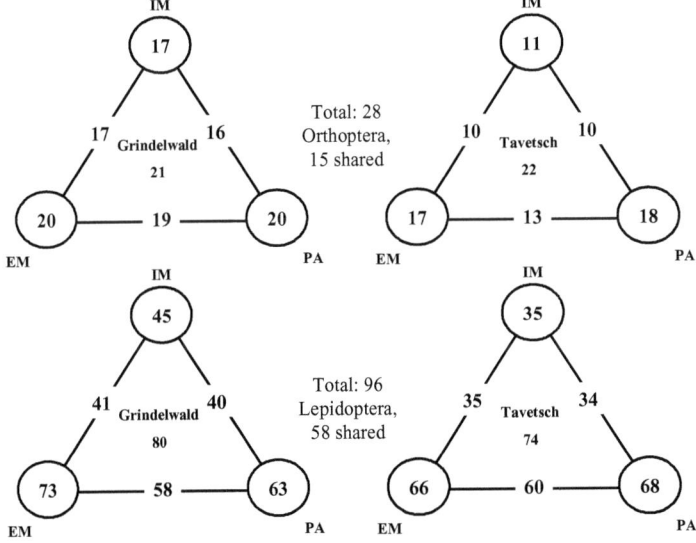

Figure 2: Total number of Orthoptera and diurnal Lepidoptera species found in the two study areas and management types. IM = intensively managed meadows; EM = extensively managed meadows; PA = pastures.

In both study areas, the lowest number of Orthoptera species, i.e. 11 resp. 17 was found in intensively managed meadows. The number of species observed in extensively managed meadows and in pastures was of about equal size (17 to 20) in the two areas (Fig. 2).

Most species present in intensively managed meadows were also present in extensively managed meadows and/or pastures.

In total, 81 Rhopalocera, 11 Hesperiidae and 9 Zygaenidae were found in the two regions (Fig. 2). Because we had to pool some species (see methods chapter) statistical analyses were based on 96 species only.

In Grindelwald, 80 species, and in Tavetsch 74 were found. 22 species occurred exclusively in Grindelwald and 16 species exclusively in the Tavetsch.

In general, the number of Lepidoptera species observed in intensively managed meadows was lower (i.e. 35 and 45) than that in extensively managed meadows and pastures (63-73). The number of species found in extensively managed meadows and in pastures was about of equal size. The largest number of species was observed in the extensively managed meadows of Grindelwald.

Average number of Orthoptera and Lepidoptera species per management type

In Grindelwald, the average number of species was highest in extensively managed meadows (13.8 ± 0.9) followed by pastures (12.3 ± 2.7) and meadows with 7.4 ± 3.8 (Fig. 3).

In Tavetsch, the average number of grasshopper species in pastures was 10.6 ± 3, i.e. nearly equal to that in extensively managed meadows with 10.3 ± 1.2 species per site. In intensively managed meadows species number was the lowest with an average of 7 ± 1.7 species per site.

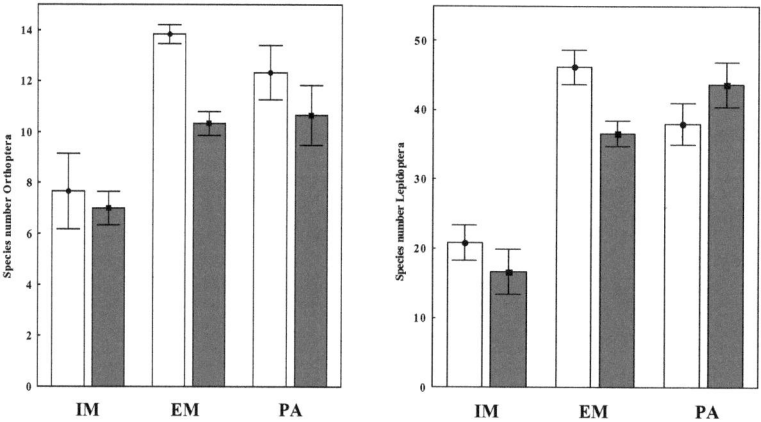

Figure 3: The average number of Orthoptera and diurnal Lepidoptera species in intensively managed meadows (IM), extensively managed meadows (EM) and lightly grazed pastures (PA) in Grindelwald (white) and Tavetsch (grey). Bars denote standard errors.

However, in both study areas, the number of grasshopper species was not significantly dependent on management type when controlling for the impacts of altitude and slope on the number of species (ANCOVA: Waldχ^2 = 0.38; ns. respectively 1.93: ns.)

In Grindelwald, the management type significantly influenced the number of Lepidoptera species even when controlled for the influence of slope and altitude (ANCOVA: Waldχ^2 = 9.68; p < 0.01). On average, it was higher in the extensively managed meadows (46.2 ± 6.4) than in pastures with 38 ± 7.7 species (Fig. 3). The average number of species in intensively managed meadows was 20.8 ± 6.6, i.e. clearly smaller than in the other management types.

In Tavetsch too, the number of Lepidoptera species was also significantly dependent on management type (ANCOVA: Waldχ^2 = 15.8; p = < 0.01). The average number of species was highest in pastures (43.6 ± 8.3) followed by extensively managed meadows with 36.6 ± 4.8 species per site (Fig. 3). In the intensively managed meadows, the mean number of species was 16.6 ± 8.3.

Table 4: The impacts of altitude, slope and management on the composition of Orthoptera and Lepidoptera communities in both study areas. Shown are the results of the Redundancy analyses on all environmental factors together and the results of the partial Redundancy analyses (pRDA) with each environmental factor separately and the covariables. df = degrees of freedom; ms = mean squares; ** = p < 0.01; * = p < 0.05.

Variable	Covariables	df	ms	% explained variance 1st axis	% total explained variance	F-ratio
Orthoptera						
Grindelwald						
Managment, altitude, slope,		5	0.10	26	49.2	3.14 **
Management	Altitude, slope	3	0.03	5.8	8.8	1.12
Altitude	Slope, management	1	0.10	9.9	9.9	2.52 *
Slope	Altitude, management	1	0.18	17.9	17.9	4.57 **
Tavetsch						
Managment, altitude, slope,		5	0.12	39	62.4	5.40 **
Management	Altitude, slope	3	0.05	10.7	15.9	2.75 **
Altitude	Slope, management	1	0.05	4.7	4.7	1.62
Slope	Altitude, management	1	0.06	5.8	5.8	2.02 *
Lepidoptera						
Grindelwald						
Managment, altitude, slope,		5	0.10	31.4	50.8	3.36 **
Management	Altitude, slope	3	0.05	9	15.2	2.01 **
Altitude	Slope, management	1	0.08	7.9	7.9	2.10 *
Slope	Altitude, management	1	0.07	7.3	7.3	1.94 *
Tavetsch						
Managment, altitude, slope,		5	0.11	32.6	55.2	4.01 **
Management	Altitude, slope	3	0.05	8.1	13.9	2.02 **
Altitude	Slope, management	1	0.09	9	9	2.60 **
Slope	Altitude, management	1	0.08	7.9	7.9	2.29 **

Composition of the Orthoptera communities

In Grindelwald, the overall Redundancy analysis of all environmental factors showed that grassland management, altitude and slope explain 49.2% of the total variance in the species data (Table 4). The Monte Carlo permutation test indicated that these factors together had a significant impact on the species composition of grasshopper communities (F-ratio = 3.14).

Figure 5 suggests the relationships between the Orthoptera species and the analysed environmental factors. Most species preferred the extensively managed meadows and pastures and were positively correlated to altitude and slope.

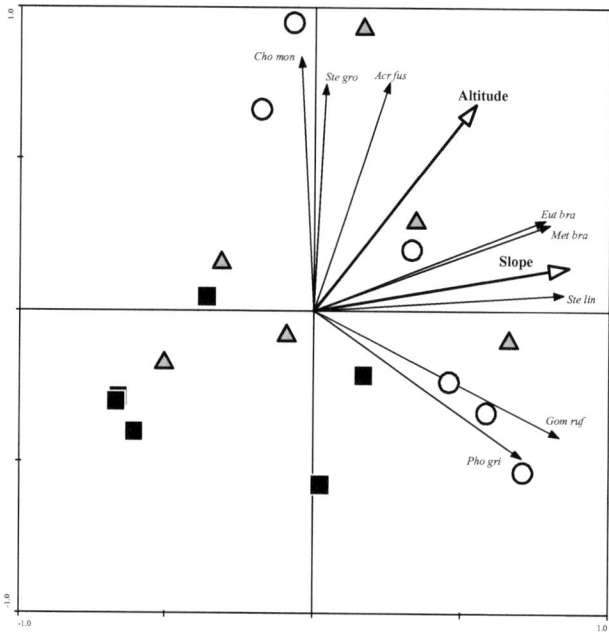

Figure 5: Ordination diagram of the Principal Component analysis (PCA) showing the Orthoptera species, the sites and their relation to the assessed environmental factors in Grindelwald. Altitude and slope are passively added in the ordination. The first two axes explain together 55.2% variance in the species data. Only the species which are highly influencing the ordination are shown, i.e. ≥ 50% of explained variance. Black squares = intensively managed meadows; Circles = extensively managed meadows; Triangles = pastures. For abbreviation of the species names see Appendix 1.

The diagram also suggests that the analysed environmental factors are correlated. Due to the possible correlations of these factors, we analysed all of them separately defining the other variables as covariables (Table 4). For these analyses we used a series of partial RDAs, which reflect the variation caused by a single environmental factor that cannot be attributed to the other factors. The decomposition of the variation (partitioning) is presented for both insect taxa and study areas in Figure 4.

The partial RDAs showed that the species composition of the grasshopper communities was not significantly dependent on management type in Grindelwald (Table 4, % total explained variance = 8.8, F-ratio = 1.12). On the other hand, the species composition changes significantly with increasing slope and altitude (F-ratio slope = 4.57; F-ratio altitude = 2.52).

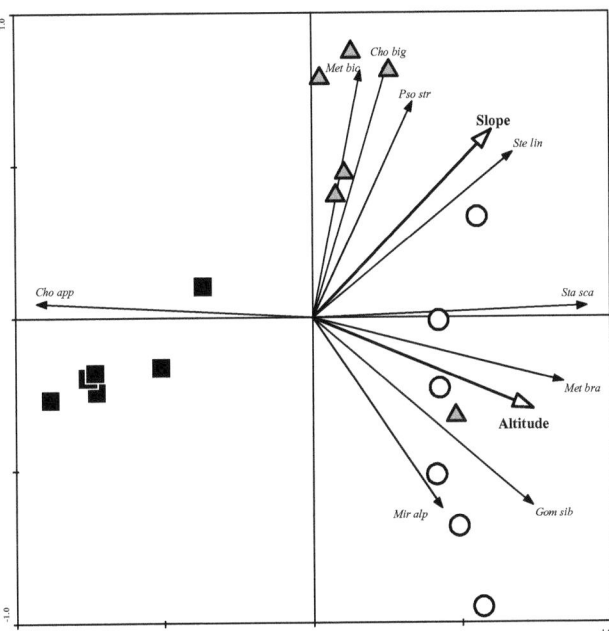

Figure 6: Ordination diagram of the Principal Component analysis (PCA) showing the Orthoptera species, the investigated sites and their relation to the assessed environmental factors in Tavetsch. Altitude and slope are passively added to the ordination. The first two axes explain together 65.4% of the variance in the species data. Only the species which are highly influencing the ordination are shown, i.e. ≥ 50% of explained variance. Black squares = intensively managed meadows; Circles = extensively managed meadows; Triangles = pastures. For abbreviation of the species names see Appendix 1.

For example, *Stenobothrus lineatus* became more abundant with increased slope in all management types (Fig. 5). Both slope and altitude explained a significant proportion of the total species data. Together they accounted for 27.8% of the explained variance (Fig. 4). Out of the two factors, slope had more impact on the species composition than altitude.

In Tavetsch, the RDA on all environmental factors showed that together they explained 62.4% of the total variance in the Orthoptera species data (Table 4). The Monte Carlo test showed that the three factors together had a significant effect on the species composition (F-ratio = 5.14).

The relation between the environmental factors and the species in the Tavetsch is shown in Figure 6. Most species preferred extensively managed meadows or pastures and were positively correlated to slope, such as *Stenobothrus lineatus*, or to altitude, like *Metrioptera brachyptera*. On the other hand, *Chorthippus apricarius* showed a clear preference for the intensively managed meadows and was negatively correlated to slope and altitude.

There were again correlations among the environmental factors; therefore, as for Grindelwald, we analysed their impact on the Orthoptera communities in the Tavetsch with partial RDAs.

The species composition of the grasshopper communities in Tavetsch depended significantly on the type of management (Table 4, % total explained variance = 15.9, F = 2.75). Additionally, the composition also changed significantly with slope (F-ration = 2.02). However, no significant change of the species composition along the altitudinal gradient could be detected.

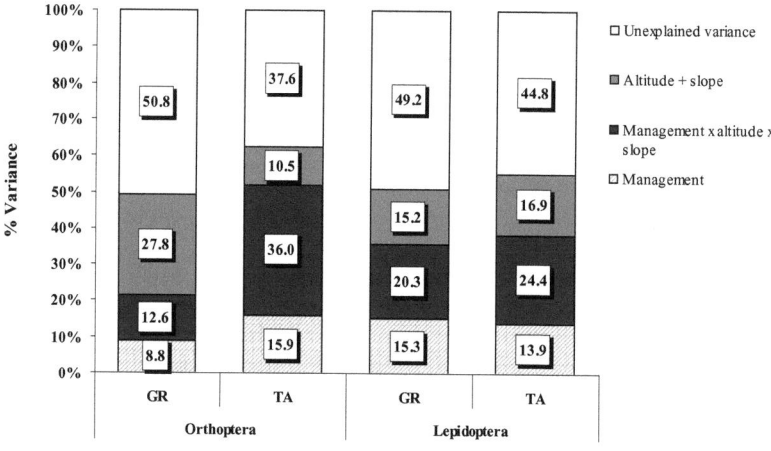

Figure 4: Variance partitioning of the species data, explained by altitude, slope and the grassland management in the two study areas. GR = Grindelwald; TA = Tavetsch.

The composition of Lepidoptera communities

The effect of grassland management altitude and slope on Lepidoptera communities was analysed in the same way as those on the Orthoptera.

The Redundancy analysis of all environmental factors and the species data in Grindelwald showed that the management, altitude and slope explain a significant proportion, i.e. 50.8% of the total variance (Table 4, F-ratio = 3.36).

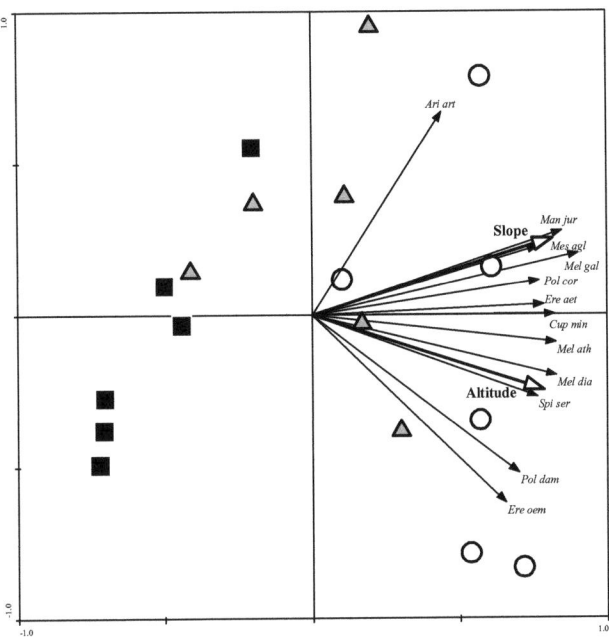

Figure 7: Ordination diagram of the Principal Component analysis (PCA) showing the Lepidoptera species, the investigated sites and their relation to the assessed environmental factors in Grindelwald. Altitude and slope are added passively in the ordination. The first two axes explain together 47.2% of the variance in the species data. Only the species which are highly influencing the ordination are shown, i.e. ≥ 60% of explained variance. Black squares = intensively managed meadows; Circles = extensively managed meadows; Triangles = pastures. For abbreviation of the species names see Appendix 1.

Figure 7 displays the relation between the assessed environmental factors and the Lepidoptera communities. Most of the Lepidoptera species showed a clear preference for extensively managed meadows and were positively related to altitude and slope, i.e. they increased in abundance with increasing altitude and slope. *Erebia oeme/E. meolans*, for example, became abundant at higher elevations while other species, such as *Maniola jurtina*, were more abundant as the sites become steeper.

The management accounted for most of the explained variance, i.e. for 15.3% (Table 4). Altitude explained 7.9% and slope 7.3%. Each of the factors had a significant impact on the composition of the Lepidoptera communities, i.e. the communities differ among the three management types and change significantly with slope and along the altitudinal gradient.

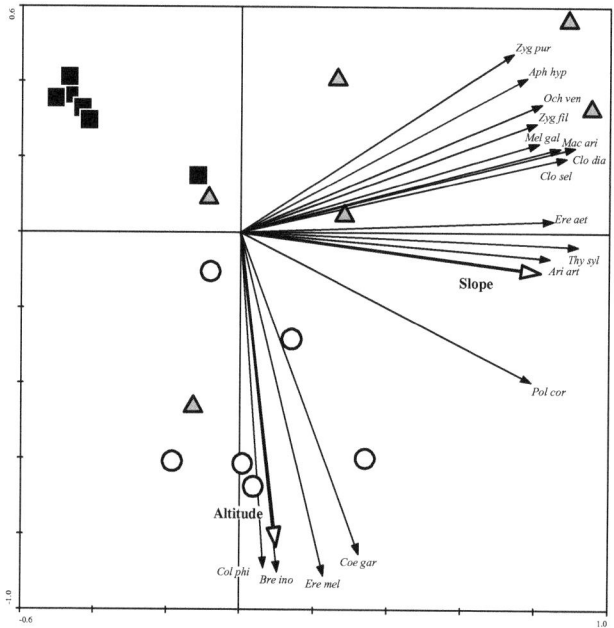

Figure 8: Ordination diagram of the Principal Component analysis (PCA) showing the Lepidoptera species, the study sites and their relation to the assessed environmental factors in Tavetsch. Altitude and slope are added passively. The first two axes explain together 57.4% of the variance in the species data. Only the species which are highly influencing the ordination are shown, i.e. ≥ 60% of explained variance. Black squares = intensively managed meadows; Circles = extensively managed meadows; Triangles = pastures. For abbreviation of the species names see Appendix 2.

In Tavetsch, the management, altitude and slope accounted for 55.2% of the total variance in the species data and the Monte Carlo test indicated that the factors had a significant impact on the species composition (Table 4, F-ratio = 4.01).

The ordination of the environmental factors and the species data of the Tavetsch showed that most species had a preference for pastures and also for extensively managed meadows and were positively correlated to altitude and slope (Fig. 8). E.g. *Polyommatus coridon* shows a preference for both pastures and extensively mown meadows and became more abundant as the sites become steeper and slightly increased with altitude. But not all species showed exactly this pattern. Species such as *Coenonympha gardetta* and *Colias phicomone* were increasing in abundance at higher elevations and preferred extensively mown grassland. Species like *Zygaena purpuralis* and *Melanargia galathea* were positively correlated to slope, but negatively to altitude, i.e. these species occurred in the steep pastures at low altitudes.

The management accounted for most of the explained variance, i.e. 13.9%, followed by altitude (9%) and slope, which explained solely 7.9% (Table 4). The Monte Carlo tests indicated that the species composition of the Lepidoptera communities differs significantly among the management types (F-ratio = 2.02) and that the composition significantly changes with increasing altitude (F-ratio = 2.6) and slope (F-ratio = 2.29).

The Orthoptera indicator species and Red List species

According to the data in Table 4, the management had no significant influence on the overall species composition of Orthoptera communities in Grindelwald. This is corroborated by the indicator species analysis concerning management types (after Dufrène & Legendre 1997). Out of the 21 Orthoptera species observed in Grindelwald, a significant indicator value (Monte Carlo permutation test; $p < 0.05$) was found only for *Euthystira brachyptera* (Table 5). This species was associated with the extensively mown meadows in this area.

In contrast to Grindelwald, in Tavetsch the grassland management had a significant overall impact on the composition of Orthoptera communities (Table 4). Thus, out of the 22 species occurring in the Tavetsch sites, for seven species a significant indicator value was found (Table 5). Two species were associated with intensively managed meadows, three with extensively managed meadows and two with pastures.

Out of the 28 species found in the two areas, eleven are on the Red List (RL) of Northern Switzerland (see Appendix 1 and Nadig & Thorens 1994). Two are rated strongly endangered (RL category 2) and seven are rated endangered (RL category 3). This RL status corresponds to that in the whole of Switzerland.

In Grindelwald, seven Red List species were observed in extensively mown meadows, seven in pastures and four in intensively mown meadows. In Tavetsch, six were found in the extensively mown meadows, six in the pastures and three in the intensively mown meadows.

The Lepidoptera indicator species and Red List species

The management type significantly influenced the species composition of the Lepidoptera communities in both areas, Grindelwald and Tavetsch (Table 4).

In Grindelwald, out of the 80 species, 18 species showed a significant indicator value (Monte Carlo permutation test; $p < 0.05$; Table 5). 16 of these species were associated with extensively managed meadows and only one each with intensively managed meadows respectively pastures.

In Tavetsch, out of the 76 species, 20 species had a significant indicator value. In contrast to Grindelwald, only five species were associated with extensively managed meadows and 15 with pastures.

In Grindelwald, *Erebia aethiops*, *Erebia meolans/E. oeme* and *Clossiana dia* were characteristic for extensively managed meadows, but in Tavetsch these species were associated with pastures. *Pieris brassicae* was associated with the intensively managed meadows in Grindelwald, but in Tavetsch with pastures.

According to Gonseth (1994), out of the 81 Rhopalocera and 11 Hesperiidae, 38 species are on the Red List of Northern Switzerland. (Grindelwald: 32 species; Tavetsch: 27 species; see Appendix 2). One species, *Pseudophilotes baton*, is critically endangered (RL category 1), five are strongly endangered (RL category 2) and 31 are rated endangered (RL category 3). This Red List status is similar to that for the whole of Switzerland.

In Grindelwald, 31 Red List species were observed in extensively mown meadows, 24 in pastures and 13 in intensively mown meadows. In Tavetsch, 22 Red List species were found in extensively mown meadows, 26 in pastures and 11 in intensively mown meadows.

Table 5: Species with a significant indicator value according to Dufrène & Legendre (1997) and their preferred management type within Grindelwald and Tavetsch. Additionally shown is the median of the abundance class of each species in the preferred management type (see Table 2, 3). Bold species are Red List species. The Red List status of Zygaenidae is unknown in Switzerland. For details see Appendix 1 and 2.

	Indicator Value %	Indicated management type	Median of abundance class in the indicated management type
Orthoptera Grindelwald			
Euthystira brachyptera	55.0	EM	4
Orthoptera Tavetsch			
Chorthippus dorsatus	84.6	IM	1-2
Chorthippus apricarius	79.2	IM	3-4
Gomphocerus sibiricus	85.7	EM	3
Metrioptera brachyptera	63.6	EM	3-4
Stauroderus scalaris	47.8	EM	4
Stenobothrus lineatus	47.9	PA	4
Chorthippus biguttulus	47.5	PA	5
Lepidoptera Grindelwald			
Pieris brassicae	48.3	IM	2
Mellicta parthenoides	83.3	EM	2-3
Mellicta athalia	80.0	EM	3
Melitaea diamina	80.0	EM	3
Zygaena loti	72.2	EM	2
Clossiana titania	69.4	EM	2
Erebia aethiops	66.7	EM	2
Erebia meolans/oeme	66.7	EM	3
Clossiana dia	66.7	EM	1
Erebia euryale adyte	60.0	EM	2
Cupido minimus	60.0	EM	2
Clossiana euphrosyne	60.0	EM	1-2
Polyommatus damon	58.0	EM	2-3
Melanargia galathea	47.5	EM	5
*Speyeria aglaia/**Fabriciana adippe** / niobe*	47.2	EM	4-5
Manjola jurtina	43.9	EM	5
Polyommatus icarus	43.6	EM	5
Lasiommata maera	60.0	PA	1-2
Lepidoptera Tavetsch			
Brenthis ino	79.2	EM	3
Colias palaeno	75.0	EM	2
Colias phicomone	69.4	EM	2
Erebia melampus	66.7	EM	5
Coenonympha gardetta	61.1	EM	4-5
Maculinea arion	76.5	PA	1-2
Lycaena phlaeas	73.1	PA	3-4
Pieris brassicae	70.6	PA	2
Erebia aethiops	65.5	PA	3
Clossiana selene	63.5	PA	2
Inachis io	62.5	PA	1-2
Parnassius apollo	60.6	PA	1-2
Clossiana dia	59.5	PA	4
Polyommatus bellargus	58.6	PA	2-3
Zygaena filipendulae	55.6	PA	1-2
Zygaena purpuralis	55.6	PA	2-3
Issoria lathonia	55.6	PA	2
Erebia meolans/oeme	53.6	PA	2
Heodes virgaurea	51.1	PA	4
Polyommatus coridon	46.8	PA	5

Discussion

Our study showed that the Orthoptera and Lepidoptera diversity in cultivated grasslands on the south slopes of Alpine valleys are influenced by management, altitude and slope. In these complex landscapes, however, the agricultural management practices and the relief are intercorrelated and thus the differences in species composition of the communities in the different management types are difficult to interpret. Hence, firstly we discuss the effects of each factor separately; then a discussion follows of their combined impacts.

The influence of management

The type of grassland management is an important factor explaining the distribution and abundance of Orthoptera and Lepidoptera. In general, diversity was higher in extensively mown meadows and in the lightly grazed pastures than in intensively mown meadows. The lowered arthropod diversity in fertilised and thus more frequently mown and disturbed grasslands matches the findings of other studies comparing intensively mown or grazed grasslands with extensively managed ones (e.g. Erhardt 1985a, b; Wingerden *et al.* 1992; Kruess & Tscharntke 2002).
Similar to the results of Erhardt (1985a, b), Dolek & Geyer (1997), Saarinen & Jantunen (2005) and others, species diversity (in the sense of species richness) of Orthoptera and Lepidoptera in extensively mown and lightly grazed grasslands was relatively high. It can be assumed that the disturbance of the vegetation by a single cut or a short period of grazing in spring and autumn have adverse effects on these two insect groups because the management lowers the availability of feeding resources and harms larvae and adults (e.g. Erhardt 1985a; Gerstmeier & Lang 1996). But the adverse effects of these two management types on the species diversity are clearly smaller than those of intensive mowing because extensively mown meadows and lightly grazed pastures remain undisturbed for a long period during the season. During this time, the majority of Orthoptera and Lepidoptera can successfully reproduce and, at the same time, these managements maintain the habitat in the long term.
The differences in the species composition of Orthoptera and Lepidoptera communities between extensively managed meadows and lightly grazed grasslands were small compared to those between these grasslands types and intensively mown meadows. Most of the species occurred in both the extensively mown meadows and in the pastures; only a few particular species clearly preferred the one or the other. Thus, both types not only harbour high species diversity, but they also represent distinct habitats that are required by particular Orthoptera and Lepidoptera species.

Interestingly, Orthoptera were less influenced by the management than Lepidoptera, particularly with regard to species number. This suggests that in the rough climatic conditions of the Alps, Orthoptera are probably less suitable indicators for the impact of grassland management than Lepidoptera. The reasons for this are manifold; the most important one is very likely the overall less pronounced association of Orthoptera to particular plant species than that of Lepidoptera. Whereas Orthoptera can choose among different grasses and herbs for nutrition (Detzel 1998; Ingrisch & Köhler 1998), Lepidoptera are often restricted to distinct host and nectar plants species (Lepidopterologen-Arbeitsgruppe 1991) which themselves are abundant only in particular management types (Ellenberg 1996).

The influence of altitude

The decline of arthropod species richness with increasing altitude is well known, indicating that cooler temperatures at higher elevations are limiting the occurrence of many taxa (Burla & Bächli 1991; Currie 1991; Wettstein & Schmid 1999). Here we showed that already an altitudinal gradient of only 600-800 m significantly influenced the species composition of Orthoptera and diurnal Lepidoptera communities in the Alps. In the Alps, no other similar studies on the change in species composition of these taxa along the altitudinal gradient are known to us. However, for moth communities, changes in species composition were observed along a transect situated at Mt. Kilimanjaro (Axmacher *et al.* 2004) at elevations between 1650 m and 3300 m, i.e. a difference in altitude of 1650 m.

In the grasslands at lower elevations, species of the colline-montane zone (e.g. *Pholidoptera griseoaptera*, among Orthoptera, see Thorens & Nadig 1997) are still present, whereas in the meadows and pastures at higher elevations subalpine-alpine species occur at the lower boarder of their distribution (e.g. species of the genus *Erebia*, among Lepidoptera, see Gonseth 1987). Thus, on the valley slopes, the Orthoptera and Lepidoptera communities represent a mixture of lowland and mountain species. This pattern partially contributes to the high species diversity on these south-facing slopes.

The influence of slope

Beside altitude, slope is another characteristic factor for the valleys. The factor also influenced the species composition of the Orthoptera and Lepidoptera communities and indicates that the diversity of both taxa is linked with the topographic variability within the south-facing slopes. Topographical variability is a major source of habitat heterogeneity (Pe'er et al. 2004) and thus leads to an increase in the overall species diversity (MacArthur 1957; May 1986; Huston 1994; Rosenzweig 1995).

In general, Orthoptera as well as Lepidoptera prefer steep grasslands probably because the microclimatic conditions in these sites are favourable to the majority of them. Species preferring dry and warm habitats, such as *Stenobothrus lineatus* (see Detzel 1998) are abundant in steep sites. We assume that most species benefit from the high radiation, high above ground temperatures (e.g. Gigon 1971) and long vegetation period on these sites, i.e. from factors that otherwise limit the development of poikilotherm organisms in the rough climate of the Alps.

On the other hand, there are a few species, such as *Stethophyma grossum*, that occurred mainly in the more flat sites. According to Detzel (1998) *S. grossum* is a typical wetland species. Such species probably benefit from the moister microclimatic conditions in the flat grasslands. In these sites, the water runoff is smaller and the soils are deeper than in steep grasslands. The coexistence of wetland species with species of dry habitats, such as *Stenobothrus lineatus*, within the same south-facing slope demonstrates well the relationships between the relief, habitat heterogeneity and species diversity in the valleys.

Combined influence of management, altitude and slope

As we have seen, grassland management and large topographic variability of the valley slopes are major sources of Orthoptera and Lepidoptera diversity in Alpine valleys. However, it is not easy to assess the influence of each single factor because they are confounded and intercorrelated. On the flat valley bottom, an intensive management is practiced because the length of the vegetation period allows at least two yields and because the sites are easy to manage. In contrast, the steep sites on the slopes are rather grazed or extensively mown. Therefore, a part of the differences in the species compositions could not be attributed to altitude and slope or to management alone, but only to their combined impact.

Furthermore, the botanical composition of the grasslands is an important additional factor influencing the occurrence of Orthoptera and Lepidoptera. In general, insect diversity (species richness) is positively correlated to plant species diversity (Erhardt 1985a; Erhardt & Thomas 1991; Vessby et al. 2001; Kampmann et al. in prep. b) and vegetation structure (Wettstein & Schmid 1999). A similar pattern was found also in our two study areas. According to Peter et al. (in prep.) in Grindelwald and in Tavetsch plant diversity was higher in the extensively mown meadows and in the pastures than in the intensively mown meadows, i.e. they found a similar pattern to the one we found with Orthoptera and Lepidoptera. But on a site basis, in general the pattern is less pronounced with the Orthoptera and Lepidoptera than that found by Peter et al. (in prep.) with the plants. An explanation of this discrepancy may lie in the mobility of these insects: within relatively short time they move from one grassland type to another, even if they are ten or hundreds of meters apart. Thus the appropriate scale assessing the species diversity (richness) of Orthoptera and Lepidoptera is not only the individual site (alpha diversity of vegetation) but also the rich mosaic of different grasslands and other habitats found in our study regions, i.e. the beta-diversity scale of vegetation.

Implications for nature conservation

In total, 30% of the Swiss Orthoptera and 46% of Lepidoptera fauna (Rhopalocera, Hesperiidae and Zygaenidae) were found in our two valleys; 38 butterflies and 11 Orthoptera are currently on the Red list of Northern Switzerland (Gonseth 1994; Nadig & Thorens 1994), stressing the high conservation value of the studied grasslands.

Our investigation documents as many others (e.g. Erhardt 1985a, b; Wettstein & Schmid 1999; Oertli *et al.* 2005) that extensive mowing and light grazing are the best grassland management forms for preserving the Orthoptera and Lepidoptera on the south slopes of Alpine valleys. In these management types the highest diversity was observed and the majority of Red List species occurred. The intensively managed grasslands, which represent the most important ones for the farmers from an economical perspective, have unfortunately, a clearly lower conservation value. However, they indirectly ensure the agricultural management of the entire region. Hence, a small scale mosaic of different management types, including intensively mown grasslands, is a key for a successful conservation of these insects.

In Switzerland, farmers get subsidies from the Swiss agri-environmental program since 1992 in order to maintain traditional and extensive management forms (Bundesblatt 1992). In the Alps, the programme helped to prevent many grasslands from abandonment (Kampmann & Herzog 2006). Nevertheless, tendencies of abandonment can be observed in both our investigation areas. Due to the economical problems in mountain agriculture, it must be expected that abandonment will further progress in future as has already happened in other parts of the Alps (Bätzing 2003). As shown by Erhardt (1985a) and Balmer & Erhardt (2000), early stages of abandoned grasslands are of high conservation value. However, in the long term, these succession stages turn into forests, i.e. habitats rather poor in Orthoptera and Lepidoptera species (e.g. Erhardt 1985a; Detzel 1998). For maintaining Lepidoptera diversity in the Swiss Jura Mountain, Balmer and Erhardt (2000) proposed a so-called rotational management, i.e. on a site, part of the grassland is managed as usual and the other part abandoned for some years, leading to different successional stages with large forbs and woody species. After some years, the abandoned part is mown or grazed again and the previously mown part is left fallow for some years. For sites that are already abandoned, they recommended a similar management with respect to different successional stages. Due to the similar landscape situation in the Alps as in the Jura Mountains, such a management scheme must also be considered in the Swiss Alps to preserve its biodiversity.

Acknowledgements

We are grateful to the farmers in Grindelwald and the Tavetsch valley for allowing us to work on their land. We thank Eugen Pleisch, Ladislaus Reser, Andreas Müller and Thomas Walter for help with identifying Orthoptera and Lepidoptera, Jessica Beller, Markus Peter and Dorothea Kampmann for helping in the field and Catherine Palmer for language corrections. The study was financed by Swiss National Science Foundation (grant 4048-064405).

References

Axmacher, J. C., Holtmann, G., Scheurmann, L., Brehm, G., Müller-Hohenstein, K. and Fiedler, K. 2004. Diversity and Distribution 10: 293-302.

Balmer, O. and Erhardt, A. 2000. Consequences of succession on extensively grazed grasslands for central European butterfly communities: Rethinking conservation practices. Conservation Biology 14: 746-757.

Bätzing, W. 1996. Landwirtschaft im Alpenraum unverzichtbar, aber zukunftslos? Pages 9-11 in Europäische Akademie Bozen, Fachbereich ‚Alpine Umwelt', editors. Landwirtschaft im Alpenraum - unverzichtbar aber zukunftslos? Eine alpenweite Bilanz. Blackwell, Wien.

Bätzing, W. 2003. Die Alpen: Geschichte und Zukunft einer europäischen Kulturlandschaft. C.H. Beck, München.

Bischof, N. 1984. Pflanzensoziologische Untersuchungen von Sukzessionen aus gemähten Magerrasen in der subalpinen Stufe der Zentralalpen. Beiträge zur geobotanischen Landesaufnahme der Schweiz 60: 1-128.

Borcard, D., Legendre, P. and Drapeau, P. 1992. Partialling out the spatial component of ecological variation. Ecology 73: 1045-1055.

Bundesblatt 1992. Botschaft zur Änderung des Landwirtschaftgesetzes vom 27. Januar 1992. Bundeskanzlei, BBL II (92.010): 1-132.

Burla, H. and Bächli, G. 1991. A search for pattern in faunistical records of drosophilid species in Switzerland. Zeitschrift für Zoologische Systematik und Evolutionsforschung 29: 176-200.

Coray, A. and Thorens, P. 2001. Heuschrecken der Schweiz; Bestimmungsschlüssel. Fauna Helvetica, CSCF and SEG, Neuchâtel, 5: 1-236.

Currie, D. J. 1991. Energy and large-scale patterns of animal and plant species richness. American Naturalist 137: 27-49.

Detzel, P. 1998. Die Heuschrecken Baden-Württembergs. Ulmer, Stuttgart.

Dietl, W. 1996. Zur pfleglichen Nutzung der Wiesen und Weiden im Berggebiet. Montagna 6: 17-18.

Dolek, M. and Geyer, A. 1997. Influence of management on butterflies of rare grassland ecosystems in Germany. Insect Conservation 1: 125-130.

Dufrène, M. and Legendre, P. 1997. Species assemblages and indicator species: the need of a flexible asymmetrical approach. Ecological Monographs 67: 345-366.

Ellenberg, H. 1996. Vegetation Mitteleuropas mit den Alpen in ökologischer, dynamischer und historischer Sicht, 5th ed., Ulmer, Stuttgart.

Erhardt, A. 1985a. Diurnal Lepidoptera: sensitive indicators of cultivated and abandoned grasslands. Journal of Applied Ecology 22: 849-861.

Erhardt, A. 1985b. Wiesen und Brachland als Lebensraum für Schmetterlinge. Birkhäuser, Basel.

Erhardt, A. 1995. Ecology and conservation of alpine Lepidoptera. Pages 258-276 in Pullin A. S., editor. Ecology and conservation of butterflies. Chapman & Hall, London.

Erhardt, A. and Thomas J. A. 1991. Lepidoptera as indicators of change in the semi-natural grasslands of lowland and upland of Europe. Pages 213-236 in Collins, N. M. and Thomas, J. A., editors. The Conservation of insects and their habitats, Academic Press, London.

Gerstmeier, R. and Lang, C. (1996) Beitrag zu Auswirkung der Mahd auf Arthropoden. Zeitschrift für Ökologie und Naturschutz, 5: 1-14.

Gigon, A. 1971. Vergleich alpiner Rasen auf Silikat- und Karbonatboden. Veröffentlichungen des Geobotanischen Institutes der Eidg. Techn. Hochschule, Stiftung Rübel, Zürich, 48: 1-164.

Gonseth, Y. 1987. Verbreitungsatlas der Tagfalter der Schweiz (Lepidoptera, Rhopalocera). Documenta Faunistica Helvetiae, CSCF, Neuchâtel, 6: 1-242.

Gonseth, Y. 1994. Rote Listen der gefährdeten Tagfalter der Schweiz. Pages 48-51 in P. Duelli, editor. Rote Listen der gefährdeten Tierarten der Schweiz. BUWAL, Bern.

Gonseth, Y., Wohlgemuth, T., Sansonnens, B. and Buttler, A. 2001. Die biogeographischen Regionen der Schweiz. Erläuterungen und Einteilungsstandard. Umwelt Materialien Nr. 137, BUWAL, Bern.

Hohl, M., Jeanneret, P., Walter, T. Lüscher, A. and Gigon, A. 2004. Spatial and temporal variation of grasshopper assemblages recorded in 1981-83 and 2002-03 in Grindelwald, Northern Swiss Alps. Grassland Science in Europe 10: 124-127.

Houston, M. A. 1994. Biological Diversity: The coexistence of species on changing landscapes. Cambridge University Press, Cambridge.

Ingrisch, S. and Köhler, G. 1998. Die Heuschrecken Mitteleuropas. Westarp Wissenschaften, Magdeburg.

Internationale Alpenschutzkomission CIPRA 2001. 2. Alpenreport: Daten, Fakten, Probleme, Lösungsansätze. Haupt, Bern, Stuttgart, Wien.

Kampmann, D. In perp. a. Dissertation an der Universität Freiburg, Freiburg.

Kampmann, D., Herzog, F., Jeanneret, P., Konold, W., Peter, M., Walter, T., Wildi, O. and Lüscher, A. In prep. b. Influence of site conditions and management type on grassland biodiversity in the Swiss Alps.

Kampmann, D. and Herzog F. 2006. Ökomassnahmen im Bergebiet erhalten die Artenvielfalt. Agrarforschung 13; 56-61.

Kiser, K. 1987. Tagaktive Grossschmetterlinge als Bioindikatoren für landwirtschaftliche Nutzflächen der Zentralschweizer Voralpen. Dissertation, Universität Freiburg Schweiz, Freiburg.

Kruess, A. and Tscharntke, T. 2002. Grazing intensity and the diversity of grasshoppers, butterflies, and trap-nesting bees and wasps. Conservation Biology 16: 1570-1580.

Lepidopterologen-Arbeitsgruppe 1991. Tagfalter und ihre Lebensräume. Vol. 1. Schweizerischer Bund für Naturschutz, Basel.

Leraut, P. J. A. 1997. Liste systématique et synonymique des Lépidoptères de France, Belgique et Corse. Suppl. à Alexanor, Paris-Wetteren.

Lüscher, A., Hohl, M., Kampmann, D., Peter, M., Herzog, F. and Jeanneret, P. 2003. Driving forces for changes in management and biodiversity of Alpine grasslands – basis for planning future development. Progress Meeting des NFP48, Progress Report, Bern.

MacArthur, R. H. 1957. Population ecology of some warblers of northeastern coniferous forests. Ecology 39: 599-619.

McCaullagh, P. and Nelder, J. A. 1989. Generalized linear models. Chapman and Hall, London.

May, R. 1986. The research for patterns in the balance of nature: advances and retreats. Ecology 67: 1115-1126.

Mueller-Dombois, D. and Ellenberg, H. 1974. Aims and methods of vegetation ecology. Wiley, New York, Chichsester, Brisbane, Toronto.

Nadig, A. and Thorens, P. 1994. Rote Liste der gefährdeten Heuschrecken der Schweiz. Pages 66-68 in Duelli, P.editor. Rote Listen der gefährdeten Tierarten der Schweiz, BUWAL, Bern.

Oertli, S., Müller, A., Steiner, D., Breitenstein, A. and Dorn, S. 2005. Cross-taxon congruence of species diversity and community similarity among three taxa in a mosaic landscape. Biological Conservation 126: 195-205.

Pe'er, G., Saltz, D., Thulke, H. H. and Motro, U. 2004. Response to topography in a hilltopping butterfly and implications for modelling nonrandom dispersal. Animal Behaviour 68: 825-839.

Peter, M. 2006. Long term floristic changes of permanent grasslands of the Swiss Alps. Dissertation ETH Zürich, Zürich.

Pfister, H. 1984. Grünlandgesellschaften, Pflanzenstandort und futterbauliche Nutzungsvarianten im montan-subalpinen Bereich. Schlussbericht des Schweizerischen MAB-Programm Nr. 7, Bundesamt für Umweltschutz, Bern.

Rosenzweig, M. L. 1995. Species diversity in space and time. Cambrige University Press, Cambridge.

Saarinen, K. and Jantunen, J. 2005. Grassland butterfly fauna under traditional animal husbandry: contrasts in diversity in mown meadows and grazed pastures. Biodiversity and Conservation 14: 3201-3213.

Schiess, H. 1988. Wildtiere in der Kulturlandschaft. Schlussbericht des Schweizerischen MAB-Programm Nr. 35, Bundesamt für Umweltschutz, Bern.

Ter Braak, C. J. F. and Prentice, I. C. 1988. A theory of gradient analysis. Advances in Ecological Research 18: 271-317.

Ter Braak, C.J.F. and Šmilauer, P. 2002. CANOCO Reference Manual and CanoDraw for Windows - User's Guide: Software for Canonical Community Ordination, Version 4.5. Biometris, Wageningen and České Budejovice.

Thorens, P. and Nadig, A. 1997. Atlas de distribution des Orthoptères de Suisse. Documenta Faunistica Helvetiae, CSCF, Neuchâtel, 16 : 1-236.

Vessby, K., Söderström, B., Glimskär, A. and Svensson, B. 2001. Species-richness correlations of six different taxa in Swedish semi-natural grasslands. Conservation Biology 16: 430-439.

Wildi, O. 1991. MULVA-4, a processing environment for vegetation analysis. Pages 407-428 in Feoli, E. and Orloci, L., editors. Computer assisted vegetation analysis. Handbook of Vegetation Science 11. Kluver Academic Publishers, Dortrecht.

Wettstein, W. and Schmid, B. 1999. Conservation of arthropod diversity in montane wetlands: effect of altitude, habitat quality and habitat fragmentation on butterflies and grasshoppers. Journal of Applied Ecology 36: 363-373.

Wingerden, W. K. R. E., Kreveld, A. R. and Bongers, W. 1992. Analysis of species composition and abundance of grasshoppers (Orth., Acrididae) in natural and fertilized grasslands. Journal of Applied Entomology 113: 138-152.

Appendix

Appendix 1: List of Orthoptera and their abundances (see method chapter). The table is ordered in a similar way as vegetation tables (Mueller-Dombois & Ellenberg 1974). The ordering was calculated with MULVA 4 (Wildi 1991) using application 5. IM = intensively managed meadows; EM = extensively managed meadows; PA = Pastures. Red List categories of Northern Switzerland (Nadig & Thorens 1994): 2 = strongly endangered; 3 = endangered.

Species	Red List category of Northern Switzerland	Management types in Grindelwald	Management types in Tavetsch
		1 2 3 4 5 6 7 8 9 10 11 12 13 14 15 16 17 18 IM4 IM1 PA2 EM1 EM5 EM4 EM3 PA3 PA5 IM6 PA1 IM3 IM2 PA6 IM5 PA4 EM2 EM6	19 20 21 22 23 24 25 26 27 28 29 30 31 32 33 34 35 36 PA6 EM5 EM3 EM1 EM6 IM3 IM2 IM4 IM6 IM1 IM5 PA1 PA3 EM4 EM2 PA4 PA5 PA2
Pholidoptera griseoaptera		2 2 4 3 4 3	
Gomphocerippus rufus		3 3 4 3 2 3 4 2	2 2
Metrioptera roeseli		4 4 5 3 2 3 4 3 3 4 4 4 3 4 3	
Acryptera fusca		1 2 2 3	2
Stethophyma grossum	2		3 3 3
Chorthippus montanus	3		3 4 2
Decticus verrucivorus	3	3 2 2 3 3 3 4 3 3 2 4 3 3 3 3 3 4	2 1 2 2 3 2 3 3 2 2 2 2 2 2
Chorthippus dorsatus		2 3 1 3 2	2 2 1 2 4 1 1 3 2 2 2 2 2
Chorthippus apricarius			2 4 4 2 3 2 2 1 2
Psophus stridulus	2		3 2 2 2 2 1
Stauroderus scalaris			4 3 4 4 1 2 2 4 2 3 4 3 3 3
Gomphocerippus sibiricus			2 4 3 4 2 3 2 3 2 1
Chothippus parallelus		4 4 3 2 3 2 4 4 5 3 4 4 4 4 4 4	4 3 4 4 3 3 3 2 3 3 3 3 3 4 4 4 3
Omocestus viridulus		2 1 3 2 3 3 3 2 5 3 2 3 3 2 3 3	3 3 3 3 3 4 3 3 2 2 2 3 4 2 2 2 3
Chorthippus biguttulus		4 3 3 3 3 4 3 2 4 2 4 3 3 2	3 3 3 3 2 2 3 2 2 3 4 4 5 5 5
Stenobothrus lineatus		3 3 4 4 3 2 4 2 1 3 2	3 3 3 3 1 2 2 2 1 1 4 5 4 3 4
Metrioptera brachyptera		1 2 3 3 2 4 4 3 2	1 2 3 4 3
Euthystira brachyptera	3	2 2 4 5 4 4 2 4 4 2 3	5 4 3 4 1 1 2 3 3 2
Tettigonia cantans		2 2 1 3 3 1 1 3 1 2	1 4 1 2 3 2
Tettigonia viridulus		1 1 1 3 1 1	1 1
Gryllus campestris	3	2 1 2 1 2 2	2
Chorthippus brunneus		2 1 2 4 2	
Metrioptera bicolor	3		3 4 3 4 2 4
Miramella alpina	3	2 1	4 3 3
Platycleis albopunctata	2	1 1 4 1	
Omocestus rufipes	3	2 1 2	

Species that were found in only one site: *Miramella formasanta* in 22 (RL category 3), *Pholidoptera aptera* in 23

Appendix 2: List of Lepidoptera and their abundances (see method chapter). The table is ordered in a similar way as vegetation tables (Mueller-Dombois & Ellenberg 1974). The ordering was calculated with MULVA 4 (Wildi 1991) using application 5. IM = intensively managed meadows; EM = extensively managed meadows; PA = Pastures. Red List categories of Northern Switzerland (Gonseth 1994): 1 = critically endangered; 2 = strongly endangered; 3 = endangered (in Switzerland no Red List for Zygaenidae exists).

Species	Red List category of Northern Switzerland	Management types in Tavetsch																	Management types in Grindelwald																	
		1	2	3	4	5	6	7	8	9	10	11	12	13	14	15	16	17	18	19	20	21	22	23	24	25	26	27	28	29	30	31	32	33	34	35	36
		EM5	PA6	EM1	EM4	EM2	EM3	IM3	IM1	IM4	IM2	IM5	PA3	PA2	PA5	PA4	PA1	EM6	IM6	EM1	EM2	PA5	EM3	EM5	EM4	PA4	EM6	PA1	PA2	IM2	IM1	PA2	PA6	IM3	IM5	IM4	IM6
Colias palaeno	3	2	2	1	2	1	2	2												1																	
Erebia tyndarus		3	1		4	2																															
Thymelicus sylvestris		2	4	2	1	3	3						5	3		5	3	3	2	5	3	3	3	5	3	3	5	2	5	2	4	5	2				
*Erebia melampus/sudetica**		5	4	3	5	5	5	1					1	1	2	1	5			3	4	4	3	2	2	2	2	3	1	2	1	1	2				
Heodes virgaureae	3	5	5	5	5	5	5		1				4	2	3	5	4	5	3	3	5	5	3	3	3	5	5	5	5	2	5	5	2		2		
Coenonympha gardetta		5	2	4	5	5	4						2	1	2	2	2	3	1	2	2	2	2	1	3	2	2		1	2		2					
Brenthis ino	3	5	5	3	3	3	3						2	1	2	2	2	1	2	3	2	2	2	2	1	2	2		1	3	2	2					
Lycaena phlaeas		2			2		1						4	4	3	3	1	4	2	2	4	2	2	2	2	5	2	2	2	2	3	2		1			
Inachis io		2	1			2	1						2	1		3	2			2	1	3	4	1	2	1	2	2	2	2	1						
Adsicta alpina/geryon		1				3	5						3	2		5				3	5	5	5	2	2	3	5	3	5	2	5	4	5	2		2	
Melanargia galathea						3	2						5	5	3	5	3	5	3	3	3	3	5	3	3	3	5	5	3	5	3	3	2	3	2		
Erynnis tages					1	1	2							1	3	3	4	2	1	3	4	2	2	3	2	2	2	3	2	4	1	1	2	1			
Thymelicus lineola			2	2		2	5						3	3	4	2	2	6	1	3	5	5	3	3	3	5	5	4	3	5	5	2	2	2			
Erebia euryale adyte			2		2	2	2		1	1	1		3	3	2	2	2	2		3	5	1	2	2	2	1	3	3	2	1	2		1		2		
Palaeochrysophanus hippothoe				2	2	2	2						2	2	2	2	2	1	5	3		1	2	2	1	3	2	2	1	2	2	1	1	2			
Hesperia comma					1	2	2		1		2		2	2	2	2	5	3	2	2	4	2	2	2	3	5	2		2	2	3	5					
Erebia aethiops	3		3		1		2						5		2	2	3	2		2	3	2	2	2	2	2	1	2	2	1							
Cupido minimus	3	5	1		1		3							1	1	2	1	1		2	3	3	2	2	2	2	2	2	2	2	1						
Polyommatus bellargus	3		3	4	2	1	4	2		2	3		4	3	3	5	2	1		3	3	4	3	2	2	1	2	4	3	2		1		1			
Polyommatus coridon	3/3	3	4	2	1	5	6	4			1	2	5	3	4	5	6	5	3	5	5	5	4	3	5	3	3	5	4		3						
Erebia meolans/oeme	3/3	2	3		3	4	3	2					2	1	2	2	5	2	1	3	2	1	1	1	1	2	2	1	3	2	5						
Clossiana dia	2				1	2	3						5	4	2	5	4	5	4	1	5	1				1	1										
Mellicta parthenoides	2																			2	2		3		3	3		3		2							
Polyommatus damon	3																			2	3	2	1	5	2	5	5				1	1					
Melitaea diamina	3																			3	3	2	5	1	2	2	2		3			2					
Spialia sertorius																				1	2	3	1	3	1	2	3		2	2		2					
Coenonympha pamphilus																				5	3	4	3	4	5	4	2	4	3	4	4	3	3	1	3	4	4
Pyrgus malvae																				4	1	1	1	3	1	2	4	1	1	1	2	3	3	2			
Zygaena loti	3				2	5												2		3	5	1	1	2	1	1	2		2								
Gynopteryx rhamni																																				1	2

Appendix 2 continued

	Red List category of Northern Switzerland	Management types in Tavetsch																		Management types in Grindelwald																		
		1	2	3	4	5	6	7	8	9	10	11	12	13	14	15	16	17	18	19	20	21	22	23	24	25	26	27	28	29	30	31	32	33	34	35	36
		EM5	PA6	EM1	EM4	EM2	EM3	IM3	IM1	IM4	IM2	IM5	PA3	PA2	PA5	PA4	PA1	EM6	IM6	EM1	EM2	PA5	EM3	EM5	EM4	PA4	EM6	PA1	PA2	IM2	IM1	PA2	PA6	IM3	IM5	IM4	IM6
Colias alfacariensis/hyale		2	2	2	2	4	2	2	5	3	3	3	3	4	3	2	4	3	3	4	4	4	3	3	3	3	4	2	3	3	4	3	3	2	3	2	3
Aglais urticae		2	3	3	1	3	5	3	3	4	3	3	5	3	3	3	3	3	3	3	4	3	5	3	4	3	3	3	4	3	2	3	3	2	3	3	2
Cynthia cardui		3	3	1	2	2	2	2	3	3	3	4	4	3	5	2	4	3	3	2	5	3	5	5	2	3	3	2	2	2	2	2	2	2	2	2	2
Polyommatus icarus		2	2	1	2	3	2	3	3	3	3	3	3	3	4	4	4	3	3	2	5	2	5	4	4	2	3	3	4	3	3	3	3	3	5	2	2
Maniola jurtina		2		3	5	3	3	1					1	5	1	2	1		1	5	5	4	5	5	4	4	3	3	3	3	3	5	4	3	1	4	
*Speyeria aglaia/ Fabriciana adippe * / niobe*	-/3/3		2	2	3		3			1		3	3	2					1	5	3	2	3	3	4	3	3	3	4	3	3	3	3	2	2		3
Aporia crataegi	3	1	1	2	1	2	5		2		2	1	3	2	2	2	2	2	3	3	3	1	3	3	2	3	3	3	2	3	2	2	2		2	2	1
Pieris rapae		1	1		2	2		2	2	2	1		2	2	2	3	2	2	5	3	2	2	2	2	3	2	2	2	2	2	2	1	3			4	2
Papilio machaon		3	2			2	3	2	3	2	2		2		2	3	2	3	2	3	3		2	3	3	2	2	3	3	2	2	2	3	3	3	3	2
Colias crocea				1		1		5	1		2					2		2			3	2	2	2	2	2	3	3		2		1		2		1	2
Pieris brassicae			1	2			5	1		1			2	3	1		2	3	2		3	2	2		2	2	2	3	2	3		2		2	3		2
Mellicta athalia	3		2	5	1	3		1	1	2	2	3	4			3	4		3	3	3	1		4		3		3		2	2		2	2	1		
Vanesa atalanta			2				1		2			1		1	1	2	2	3	3	2	2		4	5	3	2	4	5	2	3	2	2	2				
Cyaniris semiargus		1	3	1			1		2			2	2	1	2		2	2	2	2			2		3	2		3	2	3	2		2		3	2	
Heodes tityrus		2				2	2				1	2				2		2		3		2				3	2	3	2	2		2	2				
Aphantopus hyperantus			1								1			3	3	5	3	2		3	5			2	3	3		6	3	2	2	3	1		2		
Aricia artaxerxes		2	1		2	3		1	1	1	2			4	2	2		3	3	2	5	2	2	3	2	2	2	2	2		2		3			2	
Maculinea arion	3	1			1	2		1					3		1				2	1	2	2	2		3	2	2			2							
Lasiommata maera										1	1		5		2	3	3	2	2	2	2				2	2	2	2	2			2					
Zygaena transalpina		2			2	3							2	3		2	4				2			3		2			2	2	2	2					
Ochlodes venatus												1		3	5	5			3	3	2		3	2	3	2		2	2	2							
Parnassius apollo	3						1			1			1		2	2	3		2	3						2	1	2	2	3			2				
Leptidea sinapis			1	1		1							3		2		4	2	2	2	3			3	2			3	3		2	2				1	1
Zygaena filipendulae													3	1	1	4	2		3	2	3			2		2	1	2	2		1			2		1	
Issoria lathonia		1		2		1	2			1			2		2	2	2	3	2			1	1					2		1	2						
Erebia ligea													2		3		3				1		5					1					2				
Anthocharis cardamines			1	2																		1	5	5					2								
Zygaena purpuralis			1	1									4	4	5	5	1		2			1		3					2	1							
Clossiana selene	3		1		4									5	2	1		6	2		1			5													
Clossiana titania	3			2		1							2	2										5													
Callophrys rubi	3			1										1		2	2				2	1		3	2		2			1							

94

Appendix 2 continued

	Red List category for Northern Switzerland	Management types in Tavetsch																		Management types in Grindelwald																	
		1	2	3	4	5	6	7	8	9	10	11	12	13	14	15	16	17	18	19	20	21	22	23	24	25	26	27	28	29	30	31	32	33	34	35	36
		EM5	PA6	EM1	EM4	EM2	EM3	IM3	IM1	IM4	IM2	IM5	PA3	PA2	PA5	PA4	PA1	EM6	IM6	EM1	EM2	PA5	EM3	EM5	EM4	PA4	EM6	PA1	PA2	IM2	IM1	PA2	PA6	IM3	IM5	IM4	IM6
Colias phicomone		2	1		2	2	2																															
Pyrgus alveus	3						1	2									1					3			1	1	1											
Polyommatus thersites	3			1		5	1							3	1	5					1	2	1		1					2			1					
Zygaena lonicerae					2									3	1	1																	1					
Pieris bryoniae	3																				3	2		2	3		1	1								1		
Clossiana euphrosyne																					2	1		1	2						1							
Argynnis paphia																					1			2	1		1	1								1		
Adscita statices																												1										
Plebejus argus	3													5		3		2					1															
Polygonum c-album																3									2				1									
Eumedonia eumedon	3																									2		3	1				2					
Erebia manto		1			2		1																		2								2					
Zygaena viciae				2	1	1																			2													
Pyrgus malvoides				1																								1										
Mellicta varia																														1								
Polyommatus dorylas	3																																					
Eurodryas aurinia	2							2						2																								
Erebia pharte															2																							
Polyommatus eros	3																	1													1							
Hamearis lucina	3																														1							
Pararge aegeria																															1							
Lasiommata petropoliana	3																			1	1																	

Species that were found in only one site: *Apatura iris* in 30 (RL category 3), *Araschnia levana* in 12, *Carterocephalus palaemon* in 19, *Erebia medusa* in 23, *Eurodryas aurinia debilis* in 4, *Mellicta cinxia* in 19 (RL category 2), *Nymphalis antiopa* in 15 (RL category 3), *Nymphalis polychloros* in 19 (RL category 3), *Plebicula escheri* in 12 (RL category 2), **Pseudophilotes baton** in 16 (RL category 1), *Pyrgus cacaliae* and **Pyrgus serratulae** in 1 (RL category 3), *Zygaena minos* in 19

Appendix 3: Position, altitude and slope of the 36 investigated sites in the two study regions. IM = intensively managed meadows; EM = extensively managed meadows; PA = pastures

Study region	Grindelwald				Tavetsch			
	Swiss grid		Altitude	Slope	Swiss grid		Altitude	Slope
Sites	x	y	m	%	x	y	m	%
IM1	643185	165146	1070	46	702747	170598	1370	15
IM2	647223	164319	1105	16	700136	169829	1400	0
IM3	641521	164875	900	0	701667	170572	1410	0
IM4	641110	165243	920	18	699927	170025	1460	0
IM5	644525	163636	940	0	700915	170704	1460	0
IM6	644595	163538	940	0	699726	170219	1520	38
EM1	641435	165540	1005	61	702369	169821	1570	41
EM2	647118	165604	1370	38	700223	170662	1570	59
EM3	643592	165822	1440	60	699179	169559	1600	0
EM4	643411	165733	1450	54	698346	168923	1725	48
EM5	647670	165979	1470	63	698221	169133	1835	17
EM6	647139	166160	1475	30	699569	169515	1530	36
PA1	648288	164624	1210	23	697927	168399	1560	68
PA2	646307	164774	1210	27	702716	171071	1460	53
PA3	644132	165227	1220	38	702284	171123	1480	55
PA4	647237	166036	1435	35	702386	171132	1490	88
PA5	644322	166142	1525	62	698235	168605	1560	38
PA6	640933	164837	975	33	696446	167820	1760	60

Acknowledgements

My sincerest thanks go to Prof. Dr. Andreas Gigon for his scientific support, for reviewing all the manuscripts and his enormous patience during the time of my PhD thesis. All his knowledge and experience, his advice and comments substantially improved this study.

I am grateful to Dr. Philippe Jeanneret for coaching me at the AGROSCOP FAL Reckenholz. His analytical and statistical assistance and the reviews of the manuscripts were of great value.

I would like to thank PD. Dr. Andreas Erhardt for all his scientific comments on the manuscripts and his encouraging advice. Moreover, I wish to thank him for allowing me to use the great dataset he collected during his own PhD thesis in 1977-79.

Further thanks go to Prof. Dr. Peter J. Edwards for accepting spontaneously a further co-examination of my thesis.

I would like to thank Heinrich Schiess who allowed me to use the data he collected in the Grindelwald region in 1981-83.

Special thanks go to Eugen Pleisch who introduced me to the great world of butterflies and moths and who shared with me all his experience in catching, identifying and preparing Lepidoptera.

Further thanks go to Dr. Andreas Müller for providing me full access to the Entomological Collections of the ETH Zurich at any time.

I would like to thank Dorothea Kampmann and Markus Peter for all the discussions on our project and in particular for the funny time we spent together in the loneliness of the mountains and in the office.

I am grateful to:

- Dr. Andreas Lüscher and Dr. Felix Herzog for the management of the whole project.
- Thomas W. for the introduction in grasshopper identification and recording techniques.
- Catherine P. for the language corrections.
- Dieter H. for all the statistical support
- Christian R. and Christoph H. for helping me with the collection of landscape data.

Finally, I would like to thank my partner Jessica Beller for her personal support and her encouragement during the time of my investigations.

i want morebooks!

Buy your books fast and straightforward online - at one of world's fastest growing online book stores! Environmentally sound due to Print-on-Demand technologies.

Buy your books online at
www.get-morebooks.com

Kaufen Sie Ihre Bücher schnell und unkompliziert online – auf einer der am schnellsten wachsenden Buchhandelsplattformen weltweit! Dank Print-On-Demand umwelt- und ressourcenschonend produziert.

Bücher schneller online kaufen
www.morebooks.de

VDM Verlagsservicegesellschaft mbH
Heinrich-Böcking-Str. 6-8   Telefon: +49 681 3720 174   info@vdm-vsg.de
D - 66121 Saarbrücken   Telefax: +49 681 3720 1749   www.vdm-vsg.de

Printed by Books on Demand GmbH, Norderstedt / Germany